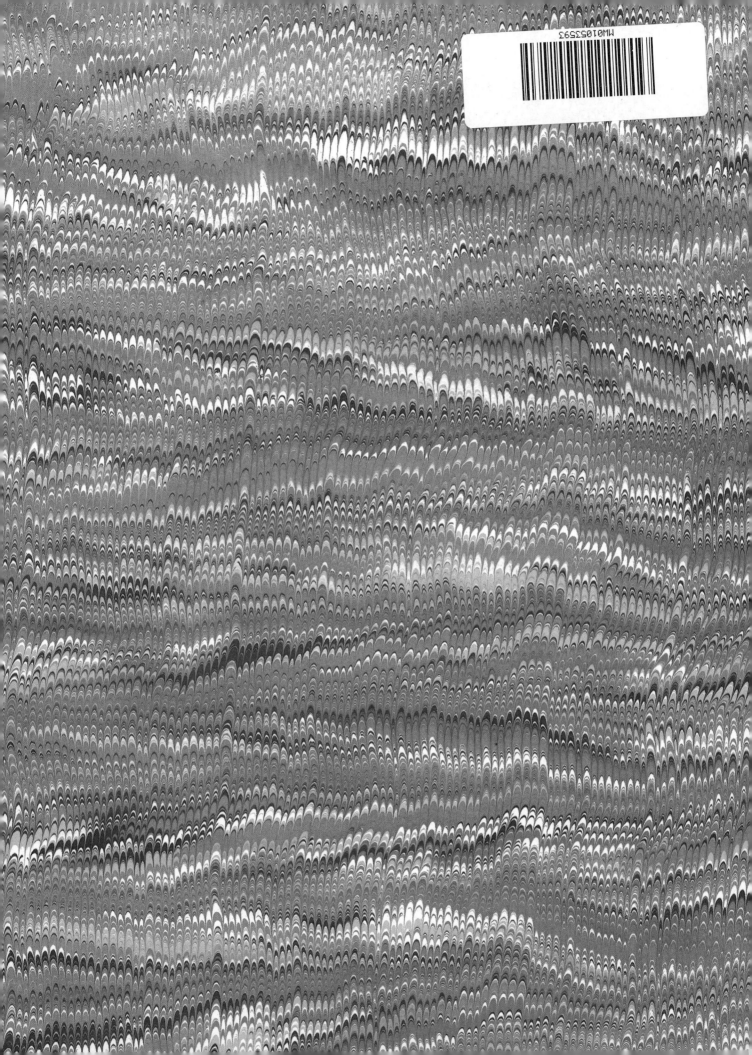

PANTHER & ITS VARIANTS

The Spielberger German Armor
& Military Vehicles Series
Vol. I

Walter J. Spielberger

Panther
& Its Variants

Schiffer Military/Aviation History
Atglen, PA

Translated from the German by Don Cox.

Copyright © 1993 by Schiffer Publishing Ltd.
Library of Congress Catalog Number: 92-60360.

All rights reserved. No part of this work may be reproduced or used in any forms or by any means — graphic, electronic or mechanical, including photocopying or information storage and retrieval systems —without written permission from the copyright holder.

Printed in the United States of America.
ISBN: 0-88740-397-2

This title was originally published under the title,
Der Panzerkampfwagen Panther und seine Abarten,
by Motorbuch Verlag, Stuttgart.

Published by Schiffer Publishing Ltd.
77 Lower Valley Road
Atglen, PA 19310
Phone: (610) 593-1777
FAX: (610) 593-2002
Email: Schifferbk@aol.com.
Please write for a free catalog.
This book may be purchased from the publisher.
Please include $2.95 postage.
Try your bookstore first.

Contents

T-34 tank (Russian) and predecessors T 46-5 (T 111), A 20, A 30 and T 32 10
Panzerkampfwagen IV — Krupp and others 11
Panzerkampfwagen VK 3001 (H) — Henschel 11
Panzerkampfwagen VK 3001 (MAN) 11
Panzerkampfwagen VK 3001 (DB) — Daimler-Benz 11
Panzerkampfwagen VK 3001 (P) — Porsche 11
Panzerkampfwagen VK 2001 (MAN) 11
Panzerkampfwagen VK 2401 (MAN) 11
Panzerkampfwagen VK 3002 (DB) — Daimler-Benz Panther prototype 11
Panzerkampfwagen VK 3002 (MAN), Panther prototype 11
Panzerkampfwagen VK 3601 (H) — Henschel 11
Panzerkampfwagen VK 4501 (H) — Henschel 11
Panzerkampfwagen VK 3002 (MAN), with Kolben-Danek steering unit (proposal) 23
Panzerkampfwagen VK 3002 (MAN), V1 with clutch-and-brake steering unit 24
Panzerkampfwagen VK 3002 (MAN), V2 with controlled differential single-radius epicyclic steering unit 26
Panzerkampfwagen VK 3002 (MAN), with OLVAR gearbox (proposal) 26
Panzerkampfwagen III — Daimler-Benz and others 24
Panzerkampfwagen Panther, Ausf. D (previously Ausf. A) — MAN, HS, MNH and DB 32
Panzerkampfwagen Panther with 75mm KwK L/100 (proposal) 29
Panzerkampfwagen Panther — Italian and Japanese programs 29, 87
Panzerkampfwagen Panther II 88, 169
Panzerkampfwagen Panther as flame thrower 89
Panzerkampfwagen Panther with hydrostatic steering unit 89
Panzerkampfwagen Panther with hydrodynamic steering unit 87
Panzerkampfwagen Panther, Ausf. A — MAN, MNH, DB 98
Panzerkampfwagen Panther with MB 507 diesel engine 108
Panzerkampfwagen Panther configured for submersion(UK) 39
Panzerkampfwagen Panther, Ausf. G — MAN, MNH, DB 125
Panzerkampfwagen Panther, Ausf. G, with all-steel resilient running gear 142, 143

Panzerkampfwagen Panther with BMW radial aircraft engine 141, 143
Panzerkampfwagen Panther with air-cooled MAN-Argus diesel engine 141
Panzerkampfwagen Panther — cost assessment 144
Panzerkampfwagen Panther with Dräger air-filtration system 146
Panzerkampfwagen Panther, Ausf. F with 75mm KwK 43 L/70 147
Panzerkampfwagen Panther, Ausf. F with 75mm KwK 44/1 L/70 150
Panzerkampfwagen Panther, Ausf. F with 88mm KwK 44 L/71 (Krupp proposal) 152
Panzerkampfwagen E 50/75 as replacement for Panther and Tiger 156
Panzerkampfwagen Panther, disguised as US Motor Carriage M 36 158
Panzerkampfwagen Panther with high-voltage cable protection 160
Panzerkampfwagen Panther in French service 160
Panzerkampfwagen Panther with infra-red night-fighting equipment 164
Infra-red observation vehicle "Uhu" 166
Infra-red escort vehicle "Falke" 169
Panzerkampfwagen Panther with electric transmission 162, 172
Panzerkampfwagen Panther with Maybach HL 234 internal combustion fuel-injected engine 174
Panzerkampfwagen Panther with air-cooled Simmering diesel engine 174
Panzerkampfwagen Panther with air-cooled Argus diesel engine and lengthened hull 174
Panzerkampfwagen 605/5 (project) 174
Aufklärungspanzer Leopard reconnaissance vehicle — MIAG 174
Auflkärungspanzer Panther reconnaissance vehicle — (project) 174
Panzerbefehlswagen Panther (Sd.Kfz.267) command vehicle 177
Panzerbefehlswagen Panther (Sd.Kfz.268) command vehicle 177
Panzerbeobachtungswagen Panther mobile observation post 183
Panzer-Selbstfahrlafette IVc 2 — Krupp 185
Panzer-Selbstfahrlafette IVd — Krupp 188
Sturmgeschütz Panther assault gun 88mm Pak 43 with

Panther II chassis (design) 193
Sturmgeschütz Panther assault gun with 88mm Pak 43/3 L/71 193
Panzerjäger Panther tank destroyer with 88mm Pak 43/3 L/71 194
Jagdpanther tank destroyer with 88mm Pak 43/3 196
Jagdpanther tank destroyer with 88mm Pak 43/4 196
Sturmgeschütz Panther assault gun with 88mm StuK 43 196
Jagdpanther tank destroyer with recoilless installation of main gun 197
Panzerjäger Panther tank destroyer with 128mm Pak 80 L/55 196
Bergepanther recovery vehicle without cable winch 200
Bergepanther recovery vehicle, Ausf. A 212
Bergepanther recovery vehicle, Ausf. G 212
Bergepanther recovery vehicle, Ausf. F (project) 214
Panther mine-clearing vehicle 214
Panther equipped with mine-clearing spade 214
Versuchsflakwagen (VFW) self-propelled anti-aircraft vehicle —Krupp 214
Versuchsflakwagen (VFW 2) Gerät 42 self-propelled anti-aircraft vehicle 214
Flakpanzer Panther anti-aircraft vehicle with 37mm Flakzwilling 341 or 44 twin guns 215
Flakpanzer Panther anti-aircraft vehicle with 55mm gun, Gerät 58 217
Flakpanzer Panther anti-aircraft vehicle with 88mm Flak 41 gun 215
Panzerhaubitze 105mm leFH 43 — Krupp 220
Panzerhaubitze 150mm sFH 43 (Grille 15) — Krupp 221
Panzrekanone 128mm K 43 (Grille 12) — Krupp 221
Panzer-Selbstfahrlafette IVb — Krupp 226
Panzerkanone Gerät 5-1213 (128mm K 43) on self-propelled chassis — Rheinmetall 227
Panzerhaubitze Gerät 5-1530 (150mm sFH 43) on self-propelled chassis — Rheinmetall 227
Panzerhaubitze Gerät 811 (150mm SFH 18/4) on self-propelled chassis 227
Panzerkanone Gerät 5-1211 (128mm K 43) on self-propelled chassis — Krupp 227
Panzerhaubitze Gerät 5-1528 (150mm sFH 43) on self-propelled chassis — Krupp 227
Ammunition carrier on shortened Panther chassis 227
Panzermörser Panther 210mm self-propelled mortar 226
Sturmmörser Panther 105mm self-propelled assault mortar 226
105mm rocket launcher on Panther chassis 228

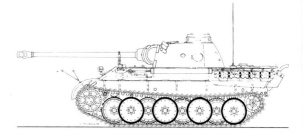

Panzerkampfwagen Panther Ausf. D, page 32 ff

Panzerkampfwagen Panther Ausf. A, page 98 ff

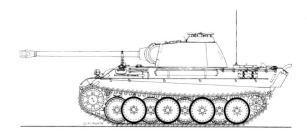

Panzerkampfwagen Panther Ausf. G, page 125 ff

Panzerkampfwagen Panther Ausf. F, page 147 ff

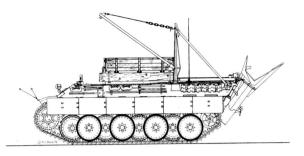

Bergepanther, page 200 ff

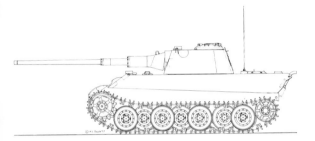

Panzerkampfwagen Panther II, pages 88, 169 ff

Flakpanzer Panther, page 214 ff

Jagdpanther, page 185 ff

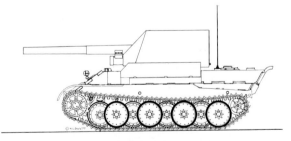

Self-propelled artillery on Panther chassis, page 220 ff

Appendixes

1 Biographical sketches of Dipl.-Ing. Kniepkamp/Dr.-Ing e.h. Maybach 230, 231
2 Compilation of data for Panzerkampfwagen Panther 232 ff
3 Compilation of data for Jagdpanther 238 ff
4 Designations for tank tracks 240
5 Surface treatment of torsion bars 241
6 Technical data 242 ff
7 Data for Panther production 244
8 Combat report of a Panther Abteilung 244 ff
9 Comparison of single-radius and dual-radius steering 246
10 Panther mockups in use 247
11 Armor thicknesses 248 ff
11a Results of shell penetration testing against armor 250

12 Panther production figures 251
13 Compilation of engagement effects against the most significant enemy tanks 252
14 Penetration effectiveness of the Panther gun 253 ff
15 Organization of a Panther tank maintenance unit 254 ff
16 Organization of a Panther Abteilung 256
17 Minutes from a meeting of the Panzer Commission 257 ff
18 Excerpts from a report by Professor Dr.-Ing. Eberan von Eberhorst 262 ff
19 Excerpts from a report by Oberst Dipl.-Ing. Esser 268 ff
20 Excerpts from the Pantherfibel 278

Index 280 ff
Bibliography 282 ff
Abbreviations 284

Foreword

It was intentional that the Panther would begin this history, for more than any other German design the Panther has exerted the greatest influence on the development of the modern combat tank.

Considering that the time span from the initial concept to the first prototype took less than a year, it can truly be said that the German Army and arms manufacturers had performed a feat unique in the history of tank development.

The Panther heavy tank was successfully employed in the field despite the many technical hurdles which go hand in hand with such an advanced design — an achievement which can be directly attributed to the untiring efforts of all those who participated in the project. It was they who found solutions in spite of the seemingly insurmountable obstacles, who often branched out into unexplored areas of technology, and who day and night had but a single goal in mind: to provide the hard-pressed troops with an outstanding combat vehicle. This they succeeded in doing in a most convincing manner. Spurred on by the Russian T-34 tank, the advanced theories of the German tank developers coalesced into a single effective solution, a solution whose influence is still being felt today. This volume attempts to once again bring to light those difficulties which faced the designers during the planning, development and manufacture of the Panther tank. The dilemma facing these individuals during this time period is demonstrated by the often conflicting interests and competence of various offices and organizations — as well as the political influence which was brought to bear.

Due to the contributions of several experts, it has been possible to give a thorough, detailed account of the long-term designs of the German tank development. This includes material ranging from the infrared battlefield illumination, the stabilized gunsight, gun stabilization, the fuel-injected internal combustion engine and the air-cooled diesel engine to the electric, hydrostatic, and hydrodynamic steering mechanism and gearbox.

I would like to express my gratitude to the following individuals for their contribution:
— Oberst a.D. Dipl.-Ing. W. Esser, the former director of the Army Testing Site for Tanks and Motorized Vehicles in Kummersdorf bei Berlin;
— Oberst a.D. Th. Icken, formerly OKH/AHA/In 6 (Inspection of Armored Troops);
— Oberstleutnant a.D. Dipl.-Ing. K. Stollberg, formerly OKH/Heereswaffenamt/WaPrüf 6 — Group director for the development of tank chassis.

I would also like to thank Peter Chamberlain, Hilary L. Doyle, Col. Robert J. Icks, Tom L. Jentz and Hofrat Dr. Friedrich Wiener for their assistance over the years.

When the Panther tank appeared on the battlefield in 1943, it was the single best combat vehicle of World War II. It was superior in terms of maneuverability, its armor protection and certainly the massive firepower of its gun. These, too, are the criteria which determine the combat effectiveness of the modern tank. With these features, the Panther epitomized the combat tank of the German army in the Second World War and paved the way for the modern vehicles of the present day. The task of this book is to document this path. We are responsive to criticism and welcome the opportunity to hear from our readers.

Walter J. Spielberger

The Panther & Its Variants

The Panzerkampfwagen IV concluded the series of tanks developed during the peacetime years. During this time, there were a number of designs submitted for heavier tanks, but none of them made it beyond the prototype stage. There was simply no one within the Army High Command who saw the necessity for a heavier combat tank.

This view changed — with a decided amount of urgency — after the appearance of the Russian T-34 in July of 1941, and work was begun on the development of heavier tanks. Like a red banner, the Russian T-34 became the standard by which to measure the tanks of the Second World War. From the German standpoint, it not only became necessary to improve upon the main guns and armor of all its tanks, there was also the need for all of Germany's antitank forces to be revamped in light of the new situation brought on by the appearance of the T-34. And finally, it was the T-34 which triggered the redesign of the Panther tank (as described in this book) and had a major influence on its development.

It is therefore appropriate to delve for a moment into the history of this Russian tank. The experience gained in the Spanish Civil War in 1937 brought about demands for a tank which was better able to withstand shell hits. The Type T 46-5 (T 111) was the first development, appearing in the spring of 1937. Following evaluations with the T 46-5 and after several interim designs, the Type A 20 appeared — the direct forerunner of the T-34. As the Model A 30, the vehicle was equipped with a high performance 76.2mm gun. Trials using the Christie running gear on the T 32 tank in 1939 and 1940 led to the Model T-34 in 1940. Making use of a reliable and powerful diesel engine, this combat tank without a doubt represented the most advanced design of the day, in spite of a simplistic clutch-and-brake steering and a primitive transmission. In a remarkable way it embodied firepower, maneuverability and armor protection. The T-34 was far superior in these areas to all western tank designs of the time. Its few weaknesses lay in its somewhat underdeveloped drive system, the limited efficiency of its running gear, the reduced command and control due to a lack of radio equipment, the inadequate visibility and in the fact that with a four man crew the commander was overtaxed. He was also the tank's gunner, and this dual role greatly reduced the fighting capability of the T-34.

Due to the thickness and angle of its armor, the German Panzertruppe and Panzerabwehr (armored and antitank forces) had an extremely difficult time combatting this altogether advanced tank design, and the shortage of adequate armor-piercing weapons was sorely felt. Even more noteworthy was the fact that the German Panzerwaffe, which were deeply impressed by the T-34, emphatically pressed for a design copy of the tank.

The following is a brief comparison of the primary construction features of the main tanks in operation at the time, including the basic data for the Panther:

	T 34 (1941)	Panzer IV Ausf. F1 (1941)	Panther (1942) Provisional Data
Combat weight (kg)	26300	22300	43000
Power-to-weight (hp/t)	19	11.9	16
Ground pressure (bar)	0.64	0.79	0.88
Muzzle velocity (armor-piercing) (m/s)	660	450	1120
Armament	7.62 cm L/30.5	7.5 cm L/24	7.5 cm L/70
Armor: front, side, rear (mm)	45/45/40	50/20+20/20	80/40/40
Fuel capacity	480 (diesel)	470 (gasoline)	720 (gasoline)
Range	455	210	240

The unpleasant surprise which the T-34 held for the German Wehrmacht could have been minimized to a certain extent. The fact that the Russians were in possession of a significantly better combat vehicle in 1941 was made known under circumstances referenced in Guderian's "Erinnerung eines Soldaten" (published in English as "Panzer Leader"). In one passage he mentions, ". . . I was quite startled, however, by an unusual event in connection with the tank in question. In the spring of 1941 Hitler had given his express permission that a Russian officer's commission be permitted to visit our tank training schools and armor production facilities, and had ordered that the Russians be allowed to see everything. During this visit, the Russians, when shown our Panzer IV, simply refused to believe that this vehicle was our heaviest tank. They repeatedly claimed that we were keeping our newest design from them, which Hitler had promised to demonstrate. The commission's insistence was so great that our manufacturers and officials in the Waffenamt finally concluded that the Russians had heavier and better types than we did. The T-34 which appeared on our front lines at the end of July 1941 revealed the new Russian design to us . . ."

All further developments were halted on the new designs for vehicles in the 30 metric ton class (VK 3001). At this time, Maschinenfabrik Augsburg-Nürnberg was at work on the VK 2001 and VK 2401 design studies as a preliminary design stage for the VK 3001. There had already been a contract awarded for a mild steel hull on the VK 3001. Of those companies working on the 30 ton project, Henschel and Porsche switched over to the development of a 36 ton and 45 ton design, respectively. The latter design would allow the acceptance of an 88 mm gun in the turret, and in doing so secure an advantage in terms of firepower.

On 18 July 1941 the firm of Rheinmetall-Borsig in Düsseldorf received a contract for the development of a tank gun with an armor penetrating power of 140 mm at 1000 meters. At the same time Rheinmetall was also given the task of designing the tank turret for the VK 3002, in which this gun was to be mounted and which later resulted in the Panther turret. A barrel with a caliber of 75 mm and a caliber length of L/60 was tested in early 1942 with performance figures nearly matching those of the requirements. This resulted in a 75 mm gun with a caliber length of L/70 planned for initial deliveries beginning in June of 1942. After thorough testing, the 75mm Kampfwagenkanone (Kwk) 42 L/70 was put into mass production. In addition to the VK 3002 turret, this gun was also planned for installation in Henschel's VK 3601 and VK 4501 turrets.

The request by the OKH/AHA/Ag K/In 6 for the development of the VK 3002 body resulted in a contract being awarded on 25 November 1941 by the Heereswaffenamt to two companies: Daimler-Benz (DB) AG in Berlin Marienfelde and Maschinenfabrik Augsburg-Nürnberg (MAN) AG in Nuremberg. The following criteria were specified for the new vehicle:

Maximum width 3150 mm, maximum height 2990 mm, ground clearance minimum of 500 mm. Engine performance was anticipated at 650-700 metric hp. An adequate cooling system was required to withstand external

The Ruscle shown here had fallen into German hands.

Reichsminister für Rüstung und Kriegsproduktion Albert Speer and Oberst Dipl.-Ing. Willi Esser, director of the Heeresversuchsstelle für Panzer und Motorisierung, seen inspecting a T-34.

temperatures of up to +42 degrees Celsius. Frontal armor was proscribed at 60 mm thickness with a 35 degree slope, while the side armor was to be 40 mm thick with a 50 degree slope. The armor protection was to be sloped over the entire vehicle. 16 mm sheet steel was considered acceptable for the floor and top of the hull. The vehicle was to have a climbing capability of up to 35 degrees and a vertical step capability of 800 mm. At a combat weight of 35 tons, the vehicle was expected to reach a speed of 4 km/h in the lowest gear and a top speed of 55 km/h. This corresponded to a maximum ratio of 13.75. The vehicle's operating endurance was mandated at 5 hours with a full load.

In view of the project's importance, discussions were held on 8 December 1941 concerning the expansion of the manufacturing facilities of MAN. The OKH insisted that the new manufacturing plant be built on soil owned by the OKH, and Röthenbach/Feucht on the Schwarzach (in Upper Franconia) was selected as the site. Construction of the facilities was to be undertaken by a company as yet to be established. MAN was to put up an initial investment of 50,000 Reichsmarks (RM). Operating capital was foreseen as 6 million RM with a yearly production expected to be 1200 combat vehicles. At an average cost per vehicle of RM 65,000, the plant would take in about 80 million RM.

Various names were suggested for the plant, ranging from a mythical figure from German legend to an honored MAN worker. References to its geographical location, such as "Schwarzach Werk GmbH", were also considered. But the name finally agreed upon was "Fränkische Fahrzeugwerke" (FFW); its first construction phase was anticipated at 57,750 square meters, costing 57.4 million RM. The monthly vehicle output would later be raised to 150 Panthers. Even though 4 million RM had already been spent for construction, all further building efforts at the Röthenbach site came to a standstill in July of 1942.

A meeting in the Heereswaffenamt on January 22nd 1942 resulted in the combat weight of the Panther being raised from its original 32.5 tons to approximately 36 tons, primarily due to changes in the construction data. A model of the new tank design was built for the Heereswaffenamt. At the same time, a model of the Daimler-Benz design was also displayed. The pleasing lines of the Daimler-Benz vehicle exhibited a rear drive, a diesel engine, a leaf spring running gear and a tapered forward hull. In order to better study the problems of a rear drive arrangement, Daimler-Benz did preliminary studies on its previously built VK 2001 (DB) using a regenerative differential steering which, through reversing the bevel gear, changed the direction of track movement. The hydraulic steering made remote operation feasible; it also initiated design studies exploring the possibility of moving the driver's seat into the turret. On 2 February 1942 Daimler-Benz received approval to submit its designs (without changes) as suitable for mass production vehicles. The first VK 3002 (DB) was to be delivered by May 1942. MAN also worked to have its first prototype ready by this time. It would have an internal combustion engine and front drive; initially, the L 600 C differential steering mechanism of the Tiger tank would be installed.

In the meantime, the Maschinenfabrik Augsburg-Nürnberg had made recommendations, accepted by the Heereswaffenamt, for the simplification of construction. Once the new form-designed steering unit had been found acceptable, the MAN solution could also be produced with a sloping forward hull. As a result, the new steering mechanism was to be made available no later than August of 1942.

A merger of the MAN design with that of Daimler-Benz was not deemed necessary, thereby dispelling the need for incorporating the Daimler-Benz diesel engine into the MAN design. MAN already possessed experience in the area of tank engine construction, having been awarded a contract by the Heereswaffenamt in 1936 for the development of an 8 cylinder diesel engine for cargo trucks. Following a few changes to this design, the 1038 G1 diesel engine was born, which provided 180 hp at 2500 rpms. One of the two test engines was installed in a Panzerkampfwagen LaS 138 and thoroughly evaluated by the Verskraft Kummersdorf. However, neither this nor the improved 1038 GL2 was pursued further.

In 1941 MAN, in cooperation with Argus-Motoren-GmbH Berlin, was awarded a contract to develop an air-cooled diesel engine for tanks. Its design was to feature 16 cylinders, type LD 220 4 cycle air-cooled diesel engine "H" type, 135 x 165mm bore and stroke, a displacement of 37.8 liters, and delivering 700 hp at 2200 rpms. Two single-cylinder engines from Argus and MAN (Augsburg) were used for testing. Trials using a full-size engine began in 1944 in Berlin and Nuremberg, but due to the war's end the evaluations were never completed.

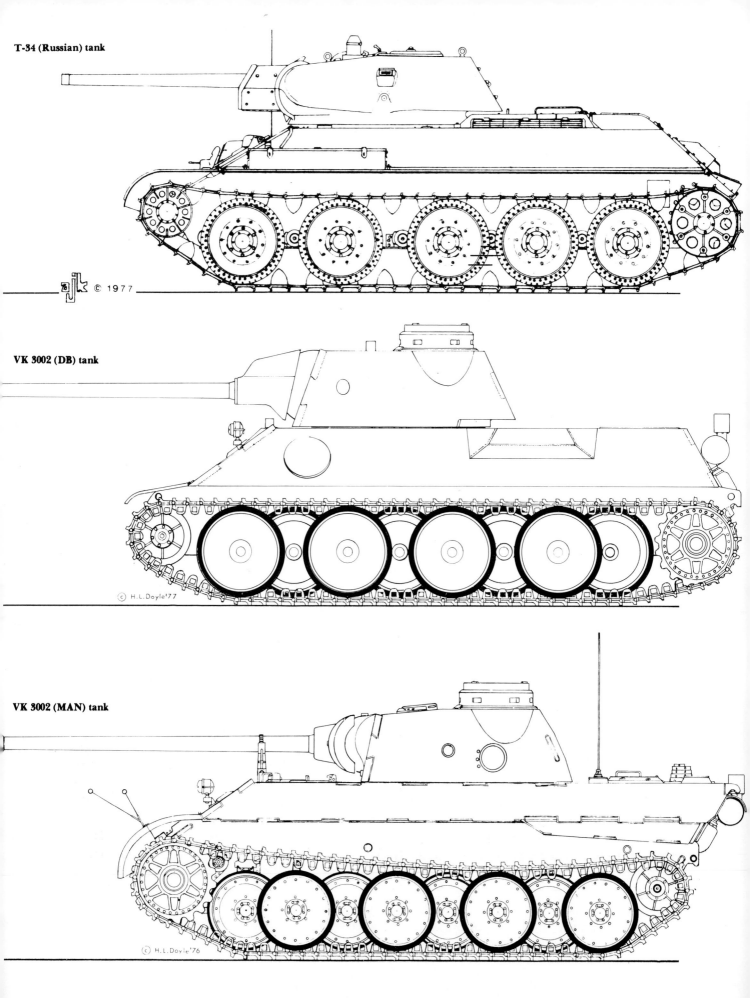

MAN was authorized to provide a liquid-cooled 650 hp diesel engine from its Augsburg factory for the Panther. A 1940 design contract from the Heereswaffenamt was used, which called for a two-stroke, 8 cylinder V-form. At 2000 rpms it was to provide 450 hp; later this requirement was increased to 650 hp. Oberst Fichtner, of the Heereswaffenamt/Wa Prüf 6, took this into consideration and recommended the performance be boosted to 700 hp if at all possible. According to data from the MAN Augsburg Werke, the first engine was expected to run in September 1942, with full production beginning a year later. Trials were only completed using a two-cylinder design; the work on the actual engine was never finished. The engine lost the interest of the developers, primarily since it was too long, too large and too heavy to be installed in the tank for which it was intended.

On the instructions of the Heereswaffenamt, the Maybach Motorenwerke in Friedrichshafen had begun development in June of 1941 on a small and lightweight 12-cylinder engine which was expected to produce 650 to 700 hp at a normal rate of 3000 rpm. In February 1942, eight months after the first pencil was laid to paper, the first individually built type HL 210 engines were delivered to Henschel and MAN. Beginning in 1943 its successor, the Type HL 230, was being produced at a rate of one engine every 25 minutes, up to 1000 engines per month. By

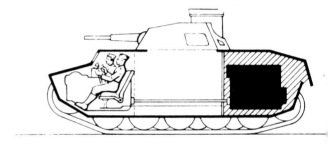

Historical development of engine compartment dimensions for German tanks 1935-1942 (Maybach).

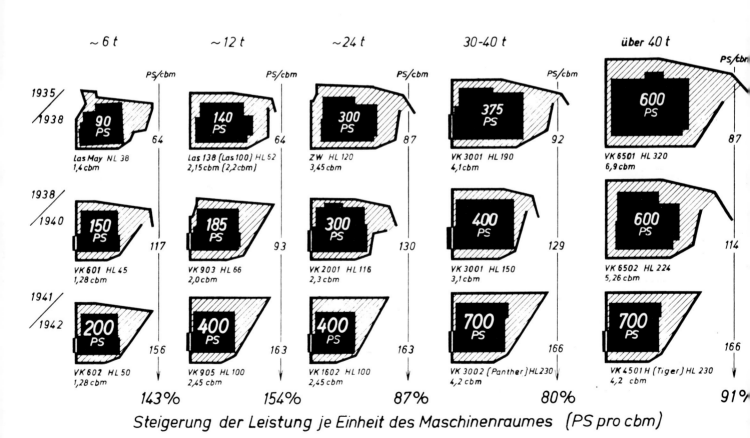

Steigerung der Leistung je Einheit des Maschinenraumes (PS pro cbm)

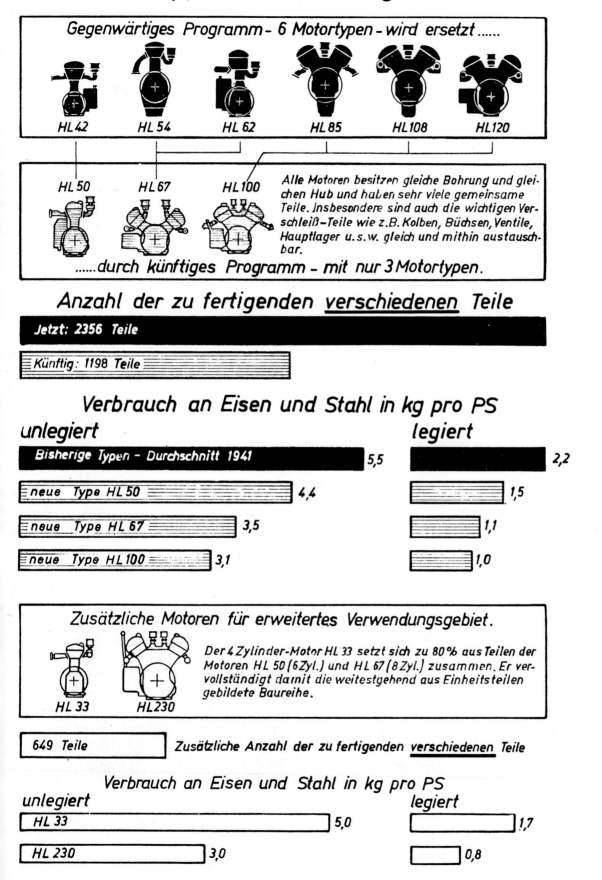

Reduction in German tank engine types, as proposed by Maybach.

the end of the war Maybach had produced 140,000 engines with a combined total of 40 million metric horsepower, and had delivered over 30,000 semi-automatic transmissions to the German Wehrmacht. Maybach was responsible for all the mechanical equipment of the tank up to the engine firewall, including among other things the dimensioning and manufacture of the cooling system, the design of the entire engine compartment, the air filtration system and for the startup equipment at low temperatures. On 3 February 1942 the Heereswaffenamt in Berlin finally established the technical requirements for

---

* Under the technical direction of Ing. Paul Max Wiebicke, born on 11 April 1886 in Chemnitz. From 1916 at MAN, Augsburg as engine designer. From 1936 technical director and responsible administrator for total vehicle development. Died on 8 March 1951 in Nuremberg.

Discussions at Maybach Motorenbau GmbH led to this study of the armor protection and running gear layout as it related to the engine length.

Aufwand für Umpanzerung und Laufwerk
auf Motorlänge bezogen

1. Motoren geringerer Abmessungen sparen an Gewicht für Umpanzerung und Laufwerk zu Gunsten größerer Waffe oder besserer Panzerung.

2. Motoren größerer Abmessungen bedingen eine Gewichtserhöhung für Umpanzerung und Laufwerk, was entweder eine Verringerung des Kampfwertes (schwächere Waffe) bzw. dünnere Panzerung zur Folge hat, oder schwerere Fahrzeuge mit wiederum größerer Antriebsleistung erfordert.

3. Größenordnung: Das Gewicht je laufenden Millimeter (l) Umpanzerung mit anteiligem Laufwerk beträgt bei den Kampfwagen je nach Stärke der Panzerung 2 bis 10 kg.

Man stelle sich vor:
2000 bis 10000 kg je laufenden Meter!

---

the Panther. Within 17 days MAN* had succeeded in clarifying all major points of the design. It was also able to obtain sufficient information on the vehicle's dimensions to make a determination as to the tooling machinery and manufacturing facilities which would be required.

Based on Speer's recommendations, on 6 March 1942 Hitler ordered that the necessary measures be directed towards mass production of the Daimler-Benz Kampfpanzer VK 3002 and to provide this company with a contract for the completion of 200 vehicles.

Hitler expected a complete report within a week and felt that the Panther design of Daimler-Benz, with its diesel engine, would prove to be superior to the internal combustion powered MAN design. He felt that Daimler-Benz would have the advantage in nearly all cases when comparing the design differences of the two types. However, the Heereswaffenamt favored the Panther design of the MAN company.

The design drawings for both projects were available by the beginning of May 1942. A Panther committee was established to evaluate these designs under the supervision of Oberst Thomale OKH/In 6 (Inspektion der Panzertruppen) and Professor Dr.-Ing. Eberan von Eberhorst, Technische Hochschule Dresden. Militarily, there were two basic prerequisites in the comparison of the two proposals:

a) The troops required a large number of this type of combat vehicle with effective armament no later than the summer of 1943. The initiation of mass production in December 1942 was therefore the first requirement. The committee felt that this requirement outweighed all others.

b) Furthermore, the troops needed weapons of superior

Daimler-Benz developed the VK 3002 to meet the Panther specifications. The photograph shows one of the prototypes with a return roller running gear. The similarity in shape to the T-34 is obvious.

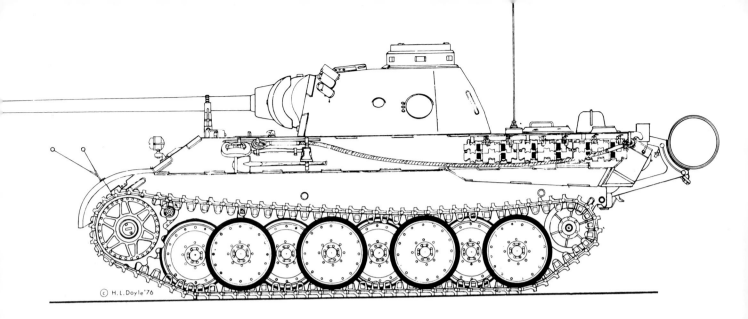

Panzerkampfwagen VK 3002 (MAN)

Wooden model of the Daimler-Benz proposal, this time with an interleaved running gear arrangement.

quality to counter the numerical material superiority of the enemy. In order to ensure that this requirement would be met, greater emphasis would have to be placed on the workmanship, rather than the most primitive solution in attaining the maximum technical performance.

Both designs fulfilled the purely tactical requirements for maneuverability, including a high combat radius with 40 km/h cruising speed on paved roads and a maximum speed of 55 km/h. The planned performance ratio of 22 hp per metric ton was not achieved by either design. The manufacturing industry would have to create an engine of greater performance, one whose installed volume and dimensions would permit it to be mounted in the Panther.

The MAN Panther would hold 750 liters of fuel, whereas the Daimler-Benz design could only carry 550 liters. The longer range gave the MAN Panther a significant advantage.

Both solutions made use of the same gun and permitted the same number of rounds to be carried. However, Daimler-Benz was not able to produce its turret in time for the deadline. In addition, its turret design required a major modification to the planned gun. The optics and machine gun mountings on the Daimler-Benz solution were more prone to shell damage than the Rheinmetall turret of the MAN Panther. The value of the attractive shape of the Daimler-Benz turret was called into question due to the fact that the diameter inside the turret was 50 mm less. This, along with other ambiguities forced the Daimler-Benz turret to be rejected. Since the Rheinmetall turret could not be mounted, it turned out that there was no acceptable turret available for the Daimler-Benz design.

Both designs met the proscribed requirements in regard to armor and surface angles. The external shape of both designs were of generally the same effectiveness. The Daimler-Benz proposal provided a greater amount of spatial comfort for the seated driver and radio operator; this was due to the use of the available rear drive, whereas MAN opted for a front drive design. However, Daimler-Benz had mounted the leaf spring suspension for the running gear externally, making the hull 100 mm narrower and reducing the diameter of the turret by 50 mm. In comparing the two designs, a debate over the advantages of a front drive vis-a-vis a rear drive was moot; time constraints required that only one design — as submitted — would be accepted. In four campaigns neither friendly nor enemy tank operations revealed any significant advantages/disadvantages to either type of drive which would cause a change in the currently accepted front drive construction. The tank's sole mission was to destroy the enemy. This it did with its weapons, not with the chassis. As a result, the function of the chassis was to support the role of the weapon in destroying the enemy.

Without a doubt, this function was best fulfilled by MAN's double torsion bar suspension system, developed by Dr.-Ing. Lehr. There was concern that there wasn't enough experience with this type of suspension, that there would be difficulties in replacing damaged or broken torsion bars, and that the manufacturing of such a system would require higher production costs. However, these concerns were outweighed by the tactical advantage of using the tank's weapons in a way which had hitherto not been possible. These concerns were overcome more easily once test vehicles and techniques had been developed which offered a maximum amount of reliability. Furthermore, there had been significant progress made in regards to the manufacture, material processing and annealing of torsion bars in comparison with the leaf spring design. The load placed on the torsion bars had been kept to a minimum by bump stops on each of three road wheel arms, thereby reducing the danger of breakage. The troops had also learned the proper way of handling and maintaining the torsion bars. In these circumstances there was the responsibility of ensuring that the superior weapon — with the potential of playing such a decisive role — which was provided to the soldier by the technician would hold up in combat. This concept clearly resulted in the valuable superiority of the German tank.

In all situations, the soldier expected an easily serviceable and reliable transmission. Neither of the designs submitted fulfilled these requirements completely.

In order to preclude any setbacks a ZF all synchro-mesh gearbox was mounted, and initially a clutch-and-brake steering was installed in the soon-to-be completed design. Although this transmission was not the ideal solution, it did provide the troops with a proven and familiar system. The two companies involved in the development of the tank were therefore able to buy the time needed to perfect their respective steering designs for installation in the vehicle in place of the original, temporary arrangement.

The amphibious qualities of the MAN design were significantly better than those of the Daimler-Benz. In comparison with the Daimler-Benz, its advantages were:
— immediate and unlimited amphibious operation
— dispensing of the necessity for operating air valves when traversing from land to water and vice versa.
— unlimited submersion time
— no loss of firing capability by the installation of the amphibious equipment

The MAN design provided for a much easier recovery of wounded personnel from the vehicle by means of hatches located immediately above the driver and radio operator positions.

With regards to maintenance and repair, the leaf spring design of the Daimler-Benz running gear enjoyed the following advantages:
— similar running gear design was already familiar to the troops
— the springs were easily replaced and maintained
— all work on the running gear could be performed outside the immediate area of the battlefield in restrictive work conditions

However, as opposed to a torsion bar suspension, replacing a center leaf spring required that all road wheels be removed. The troops still had no experience whatsoever with MAN's double torsion bar suspension system. It possessed no type of internal shock absorption and required a particularly careful touch during installation and removal. This proved to be a more difficult task than with the leaf spring design. Nevertheless, the double torsion bar suspension was conditionally acceptable for field operations if:
— torsion bar breakage was kept to a minimum

— the time required for torsion bar replacement was kept to an acceptable level

Both designs called for the installation of the 700 hp Maybach HL 230 engine. The engine performance met the requirements of the construction directives requiring a performance/weight ratio of 20 hp per metric ton, but did not meet the latest troop requirements of at least 22 hp/ton.

According to data from Maybach, the gasoline engine planned by MAN provided more favorable engine torque than previous Maybach engines, even at lower rpms.

Daimler-Benz mounted a radiator laterally on both the right and left side and drew the cooling air in behind the radiators and over the engine, exposing the engine to dust as with earlier designs.

Alternately, MAN arranged two vertically positioned radiators on each side of the engine in a special waterproof compartment which was completely blocked off from the engine itself, including the fans. The MAN design was therefore superior with regards to engine grime and submersibility. MAN dispensed with a V-belt drive for the fans through the use of bevel gears in conjunction with drive shafts. Daimler-Benz chose to drive one fan directly and the other with three V-belts.

Neither an internal combustion nor a diesel-powered air-cooled engine was available in sufficient quantities at the time. The rear drive of Daimler-Benz required a lateral arrangement of the gearbox next to the engine, which made a spur gear necessary. The hydraulically-operated main clutch was located between the spur gear and the gearbox.

The gearbox was a development of Daimler-Benz in conjunction with the firm of Ortlinghaus and utilized a hydraulic multiple disk clutch to transfer the driving power. Although this type of gear system made smooth shifting possible — even in rapid succession — and its operation appeared to be quite simple, an immediate introduction onto the assembly line was not yet possible. There was simply no experience with its suitability in vehicle operations (with the exception of small switcher diesel locomotives). A further disadvantage was its overall length.

The development of this type of transmission system would be pursued further. It was certainly possible in terms of space to install the ZF synchro-mesh gearbox, but a modification of the complete transmission system up to the drive sprocket would be required due to the distance involved. As with MAN, Daimler-Benz was therefore also required to make use of a two-stage reduction in the final drive.

The shifting of the rear-mounted gear unit required an elaborate system of transmission linkage rods, which was rejected in view of its lack of simple operation and the general design approach. Based on its simple design, the synchro-mesh gearbox was accepted in spite of the fact that certain technical disadvantages accompanied the installation of this type of gearbox in such a heavy vehicle with its naturally high rolling resistance. Due to their unproven stability the introduction of the Maybach-OLVAR gearbox, the electromagnetic ZF drive or the pulse type gearbox was still considered to be premature at this time. However, in order to facilitate driver training and ease the burden on the driver it was necessary to accelerate the testing of such transmissions prior to adopting them. The complete inter-changeability of the gearbox would be a prerequisite, which according to MAN sources had already occurred with the Maybach-OLVAR gearbox then being used in the Tiger tank.

The designs of both firms envisaged a clutch-and-brake steering. Daimler-Benz promoted a hydraulic multi-plate clutch whereas MAN's offering was a single-plate dry clutch.

TRegarding the steering brakes, there had been no experience with armored vehicles of this size and required maximum speed. It was, however, known from experience with vehicles of similar size (but slower speeds) that the brake wear and the strength required of the driver were both extraordinarily great. The parts which were subject to wear would have to be of generous size, easily replaced and able to cool quickly. Brake adjustment would have to be accomplished without leaving the vehicle and with simple tools. MAN's planned solid-disk brake was considered by the Panther Kommission as inadequate for a friction brake at that time due to thermal reasons. The controlled differential regenerative steering, intended by MAN as a further improvement, was generally felt to be more advantageous since it required less strength to operate, caused less loss of control and none of its parts appeared to be subject to a high degree of wear. In order for the tank to be able to pivot while at a stop, the commission additionally required corresponding brake steering. A regenerative controlled steering unit (with two radii of turn in each gear) had been developed for the Tiger tank and had demonstrated its effectiveness without any major teething troubles. However, the manufacturing industry maintained that the machinery for gear production — primarily those necessary for the internal teeth gearing — were not available for large-scale production of the Panther tank and could not be obtained in sufficient quantities. As an interim measure a clutch-and-brake steering mechanism was considered acceptable. Development and testing of the controlled differential steering mechanism would have to be accelerated in light of the advantages it offered. The MAN design already provided for the possibility of its installation.

Both designs took into consideration the cruising and maximum proscribed speeds and offered interleaved suspensions with eight road wheels on each side. Both proposals bordered on the maximum permissible road wheel load, yet damage to the road wheel rubber proved to be much less than feared initially. The speeds which the design called for, coupled with the weight of the vehicle, simply left no other option for a running gear arrangement. MAN had mounted independently supported bogie wheels on the swing arms while Daimler-Benz fixed the road wheels in pairs to counterbalances which pivoted on rocker arms. MAN's road wheels were cushioned individually by a double torsion bar spring. Daimler-Benz envisioned a pivoting short leaf spring for the front and rear road wheel pairs. Support for the two middle pairs was achieved by a long leaf spring.

In spite of having the same vehicle width, Daimler-Benz had a hull 100mm and tracks 120mm narrower as a result of the externally-mounted leaf spring design and its counterbalance arms. The arrangement of the Daimler-Benz running gear resulted in a higher specific ground pressure. It amounted to 0.83 bar as opposed to 0.68 bar of the MAN vehicle, which was closer to the 0.66 bar of the Russian T-34.

---

* The ingenious idea of using an interleaved running gear, which was also in use for all German armored half-tracks, stemmed from Ministerialrat Dipl.-Ing. Kniepkamp of the Heereswaffenamt/WaPrüf 6. With the given track-on-ground length, there was as much road wheel rubber laid to the track as corresponded to the rubber's low compression potential of the time.

The non-cushioned torsion bar suspension of MAN necessitated the use of shock absorbers which were mounted on the swing arms inside the tank hull of the second and second from last road wheels. It was planned to utilize the HT 90 shock absorbers of the Hemscheidt firm, which had already demonstrated their functionality with the Tiger tank.

Since the HT 90 shock absorber as mounted in the Panther was only under half the stress as the smaller Hemscheidt shock absorber of the Panzer III — thanks to the reduced natural pitching frequency due to the double torsion suspension — it was considered that its installation in the tank's interior would produce favorable results.

The most significant advantage of MAN's double torsion bar suspension was its large travel stroke (510 mm) with minimal stress being placed on the suspension system itself. After the endurance trials conducted by MAN it was found that the stress in operation (+/-16 kg/mm2 = +/-160 N/mm2) was lower than the anticipated permissible fatigue limit of +/-20 to 22 kg/mm2 (200 to 220 N/mm2) which had been derived through either surface pressurization finishing or bombarding the torsion bar surface with steel shot. The elasticity of the double torsion bars enabled unusually low natural frequencies of the vehicle in bump and rebound undulations.

The spring resilience made possible a higher cross-country speed by offering a greater capability for target tracking and by tiring the crew to a lesser degree. The more cushioned suspension also reduced the wear upon the tire rubber and the shock absorbers as well as lessening the stress on the entire running gear.

The torsion bars' potential for breakage was kept to a minimum by a deflection limiter/bump stop on the first, second and seventh road wheel arm. The resilience of suspension, which could play a decisive role in a vehicle's combat effectiveness, was not available in the leaf spring running gear of the Daimler-Benz proposal. The more advanced double torsion bar suspension was therefore given priority, even though certain maintenance difficulties would have to be taken into account.

Daimler-Benz incorporated amphibious features into its vehicle by not only sealing all hatches and covers, but also closing off the air inlets and outlets to the engine compartment from outside of the vehicle by means of valves. After the water obstacle had been crossed, these valves could then be reopened from inside the tank. This meant that the engine was not cooled during amphibious operations. Trials in the field showed that in this condition engine damage set in after only ten minutes' travel time.

As opposed to the Daimler-Benz vehicle, the radiators and airflow of the MAN tank were not sealed off. Just prior to beginning submersed operations, the fan drive is shut down from the driver's seat, thus ensuring that normal cooling occurs until actually entering the water. During submerged operations water is circulated around the radiators, enabling underwater travel of unlimited duration. This flooding of the cooling apparatus required that the MAN design be impervious to sea water.

The center of gravity was directly above the center of the track supports on the Daimler-Benz proposal, whereas MAN's design had moved it 120mm behind the center. Daimler-Benz offered a greater overlapping of its road wheels than did MAN. For reasons of weight conservation, studies would be conducted to determine whether MAN could also produce its vehicle with the same amount of overlap as Daimler-Benz.

The five-hour operational capability at full combat weight proscribed in the construction directives mandated an hourly fuel consumption rate of 240 liters. Since the 30 km/h average speed did not require full engine power, however, calculations based on experience data of the Kraftfahrtversuchsstelle Kummersdorf were taken as the new basis. Namely, these were 8 liters per vehicle ton per 100 km (8 l/t 100 km) during road travel and 11 l/t 100 km for traversing moderate terrain using gasoline fuel with an octane rating of 74. The varied fuel capacity of the two proposals — MAN = 750 liters, Daimler-Benz = 550 liters — resulted in the following statistics:

	MAN	Daimler-Benz
on road surfaces	270 km	195 km
on terrain	195 km	140 km

Advantages of the front and rear drive systems were determined to be as follows:

1) MAN front drive
— direct operation of the gearbox
— direct operation of the steering unit
— adjustment of the steering brakes possible without leaving the vehicle
— horizontal entry hatch considered more favorable
— better self-cleaning tracks (front drive was actually found to be more effective in mud and slush for cleaning the open link tracks of the time as the drive sprocket meshed with the track.)

2) Daimler-Benz rear drive
— elimination of heat, noise and odor in the fighting compartment caused by the transmission as well as the braking mechanism
— unrestricted seating room for the driver and radio operator
— more efficient use of space in fighting compartment
— lower overall vehicle height

When determining the fighting compartment area MAN did not figure in the space required for the torsion suspension which occupied the entire length and width of the vehicle. Furthermore, a cylindrical housing of the same diameter as the steering brakes was also not calculated. Then there was the space necessary for the drive and transmission systems as well as the area for the drive shaft tunnel (having a cross section of 250mm x 250mm) running the entire length from the transmission to the engine bulkhead. The revolving floor beneath the turret was situated above the drive shaft and the turret drive. These features necessitated the large height of the Panther tank. In addition, the driver's seat could not be positioned between the torsion bars.

Daimler-Benz did not make any subtractions; the entire interior hull space in front of the engine bulkhead was calculated as total area for the fighting compartment.

The MAN proposal had 7.26 square meters of fighting compartment area available, whereas the Daimler-Benz solution offered 6.43 square meters. The Daimler-Benz design had a 52 mm lower hull height and a 195 mm lower total height.

With regards to production requirements in terms of hours needed, it turned out that the Daimler-Benz chassis required approximately the same amount of production time as the MAN version in its simplified form as currently submitted. This was explained by the fact that the firm of MAN had made constructive simplifications such as replacing the regenerative controlled differential discontinuous steering with a clutch-and-brake steering and the planetary type reduction gear in the final drive with a two stage spur wheel gear. Daimler-Benz, however, had switched from its simple support roller running gear with steel road wheels to an interleaved suspension in order to meet the speed requirements. It was generally felt that the machinery then being used in manufacturing the Panzer III would, with a limited expansion program, be adequate for the simplified design of MAN as well as for the Daimler-Benz Panther. Turning machines would have to be supplied to a few companies in order to produce the turret opening in the vehicle hull. A special type of hull drill press for manufacturing the hull of the MAN vehicle was designed by MAN; ten of these would be made available.

Neither design, with their clutch-and-brake steering and interleaved running gear enjoyed a particular advantage in terms of production engineering and supply. For comparison purposes the production time of the Daimler-Benz chassis was 1063 working hours and that of MAN 1078.5.

On 11 May 1942 the committee evaluating the Panther tank, which had been meeting in the OKH Berlin, Bendlerstrasse building, gave its selection decision to the chairman of the Tank Commission, Professor Dr.-Ing. h.c. F. Porsche with the following words: "The committee evaluating the designs for the Panther tank as submitted by the firms Daimler-Benz AG and Maschinenfabrik Augsburg-Nürnberg AG (MAN), following the meetings of 1, 5, 6, and 7 May, unanimously favors the proposal of the firm of MAN in the version of an 8 roller interleaved running gear and double torsion bar suspension, ZF all synchro-mesh gearbox and clutch-and-brake steering, and recommends that the Panzertruppe be equipped with the selected tank."

This recommendation was presented to Hitler on 13 May 1942 and thoroughly discussed. Hitler felt that the armor was too weak and considered that the rear drive of the Daimler-Benz version was the best. He realized, however, that quick production could be decisive and that under no circumstances could two different types be successfully produced side by side. He agreed to study the recommendation overnight and relay his decision the next day through his adjutant Major Engel. On 14 May 1942, Major Engel reported that Hitler was in agreement with the recommendation and the MAN Panther would be built. The company was, however, to give heed to the suggestions of Professor Dr.-Ing Porsche. He had recommended that the possibility be explored of installing a Kolben-Danek steering unit, similar to that of the Pzkpfw 38 (t). In addition the glacis plate was to be 80 mm thick.

As early as 4 May 1942 a conference had been held at MAN regarding the production of the Panther (VK 3002), which resulted in the following basic data:
— internal combustion engine, Maybach HL 210/230, approx. 600 metric hp
— Maybach OLVAR gear drive 0640 1216 (as on the Tiger tank)
— fuel capacity approx. 700 liters
— spur gear side transmission doubly geared down, with sprockets of module 9 and 11.

The middle tooth group was not required to be ground since it made no contact.
— combat weight of 35 metric tons
— total vehicle length with gun of 8625 mm
— total chassis length of 6839 mm
— total chassis width of 3270 mm
— height of hull at 1314 mm
— maximum speed of 55 km/h

The hinged tracks consisted of 86 links. According to MAN, the width of the tank would not prevent its transport by the Deutsche Reichsbahn.

At this time, it was still not certain what steering type would be utilized, but it was assumed that a simple clutch-and-brake steering unit would be installed initially, since the companies making the vehicle's components did not have the proper gear cutting machines needed to produce a controlled differential steering unit (29 gear wheels per unit). There was an acute shortage of slotting machines necessary for the internal serration of the hollow gear. The housing was to be manufactured of cast steel with a strength of 60 kg/mm. Shrink holes which appeared might be welded over on the condition that the housing be annealed following the welding. In a conference with the Reichsminister für Rüstung und Kriegsproduktion (Armament and War Production) on the 19th of May 1942 it was agreed that the majority of machinery materials for the Panther would be acquired in France. Development work on the Panther of Daimler-Benz was halted. The two diesel engine prototypes under construction would be completed for study purposes; one of them would be constructed with support rollers. At the same time, Daimler-Benz was to use all available resources in rapidly gearing up for productionof the MAN design. MAN, Nuremberg, and Daimler-Benz, Berlin-Marienfelde, were to begin production that very year on the MAN Panther. The firms of Maschinenfabrik Niedersachsen-Hannover (MNH) and Henschel, Kassel, were to plan for production beginning in July 1943 in the following quantities: July — 1, August — 3, September — 5, October — 10, and November — 15; thereafter monthly production was set at 50 tanks. According to rough estimates the labor hour relation in comparison to the Panzer III stood at approximately 1 to 1.25, i.e. 4 Panther vehicles for every 5 Panzer III tanks built. Cost (without weaponry) PzKpfw III RM 96,163.00; Panther RM 117,100.00.

There was little satisfaction on June 4th, 1942 in the Reichsministerium für Rüstung und Kriegsproduktion with regards to the planned production run of the Panther. It was necessary to ensure that by 12 May 1943 there would be at least 250 Panthers available. Hitler expressed his doubts that even the thicker 80 mm frontal armor would be adequate for the spring of 1943 and demanded that at least all vertical front surfaces be protected by 100 mm thick armor plating.

On 13 June 1942 Henschel was assigned the task of setting up production of 70 Panthers per month — without affecting the output of the Panzer III then being manufactured.

A study was begun on June 18th 1942 to determine whether 254 Panthers could actually be produced by April of 1943. Daimler-Benz, Berlin-Marienfelde, was expected to have an output of 91 vehicles, Henschel, Kassel 36, MNH, Hannover 61, and MAN, Nuremberg 84 vehicles. MAN requested the transfer of a facility of the Südeisen firm in Nuremberg for production of the Panther. The Panther manufacturers recommended a consolidation of the steering mechanism production. MAN had contracted out the standard parts for 2000 vehicles, with further contracts planned for 1000 and 1500 respectively. The companies wanted to reserve the initial vehicles for testing and as models, a request which was denied. The test vehicles anticipated for August and September 1942 would be immediately needed for testing at the Verskraft

---

* At several points in this book it is mentioned that Hitler made many fundamental decisions himself which had a major effect in the area of tank development and production. In this he was decisively influenced in many cases by Hauptdienstleiter (Staatssekretär) Dipl.-Ing. Karl Otto Saur (1902-1966), director of the Technisches Amt in the Reichsministerium für Rüstung und Kriegsproduktion. Saur, who until 1937 had been with the Thyssen company, was responsible under Speer for the development and production program. He enjoyed the complete trust of Hitler and Speer and was appointed by Hitler as Speer's successor in April of 1945. Saur set himself at odds with the industry and the Heereswaffenamt with his push for an increase in war production. His understanding for thorough testing of newly developed equipment and their durability was somewhat lacking. Material quality and replacement parts took second place. Nevertheless, the growth in tank production during the war's final years without a doubt was to his credit.

in Kummersdorf and at MAN's development branch. Panther production was placed in the same category of urgency as oil and locomotive production programs. Demands were placed upon Henschel to step up its operations to where one vehicle would be completed by January 1943, five in February, in March 10 and in April 1943 20 vehicles would be delivered. Greater pressure was brought to bear on all the firms involved in order to unconditionally meet the demands of Hitler. MAN devoted its efforts to having the first test vehicle ready by August 1942 and begin mass production in November. Henschel felt that its Panther contract for 200 vehicles was too low and requested an increase to 1000 Panthers. A reduction in Panzer III production was seen as unavoidable.

The following delivery schedule was formulated:

Company	Nov. 42	Dec. 42	Jan. 43	Feb. 43	Mar. 43	Apr. 43
MAN	1	3	10	20	25	25
Daimler-Benz	-	1	10	20	30	30
MNH	-	-	1	10	20	30
Henschel	-	-	1	4	9	12
Total	1	4	22	54	84	97

The suggestion was made to equip at least 250 vehicles with the MAN clutch-and-brake steering unit until production of the single-radius epicyclic steering was assured. Henschel confirmed on 20 June 1942 that it would be able to deliver 26 tanks by the end of April 1943, but indicated that this would mean a drop in Panzer III production by approximately 100 vehicles. Henschel's recommendation was based on an increase in Tiger production to 50 tanks per month, which would lower Panzer III output by only 75 vehicles. Due to the narrow confines in the Kassel-Mittelfeld plant, it was not possible to meet both Tiger and Panther production schedules at the same time. Henschel stated that it would be impossible to continue further manufacture of Panthers without first terminating Panzer III production.

The elevation and traverse controls for the weapons systems were to be manufactured by the Lohmann Werke's branch at Pabianice near Litzmannstadt. On July 7th, 1942 a contract was awarded to produce 150 pieces monthly.

On 2 July 1942 the chairman of the Panzerkommission, Professor Dr.-Ing. h.c. Porsche made the announcement at the commission's meeting that Hitler had once again demanded that development of air-cooled diesel engines for all types of army vehicles be given top priority. Their introduction was to begin by 1943. The firms of Daimler-Benz, Klöckner-Humboldt-Deutz, Krupp, Maybach, Osterreichische.

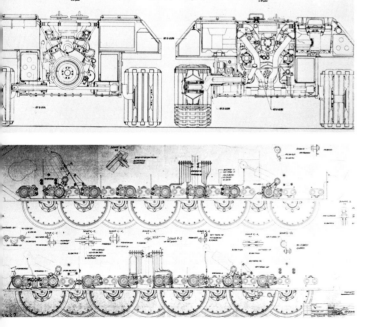
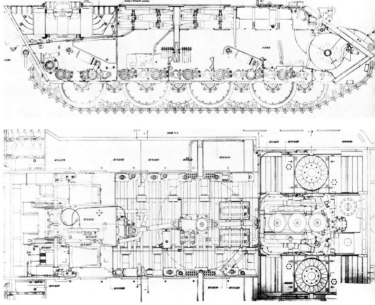

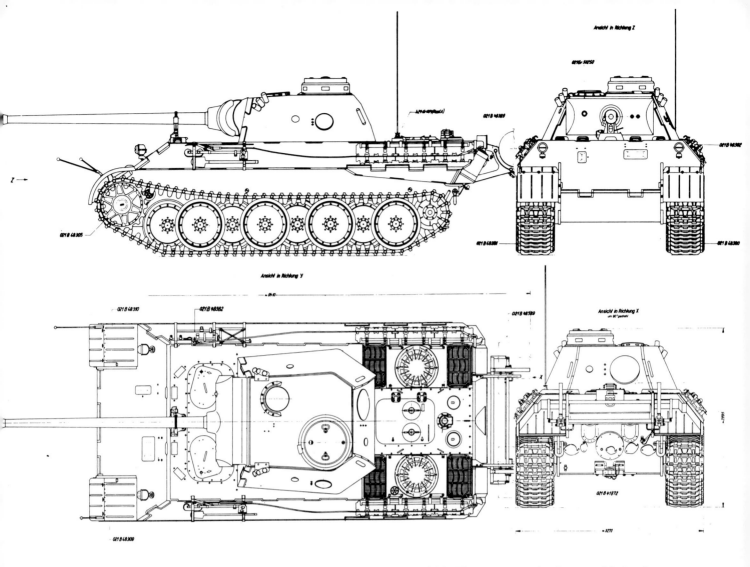

The MAN proposal already showed the form which was carried over to the production vehicle. The arrangement for the external fuel tanks on the rear of the vehicle is noteworthy, as is the fact that the tank is identified as an Ausführung A — even though the first production vehicle carried the designation Ausführung D.

Automobil Fabrik, Opel, Phänomen, Tatra, Simmering, Steyr and Wigru-Fahrzeug-Industrie were tasked with forming a work committee to share their experience and ideas for development in this matter. Oberstleutnant Holzhäuer was assigned the responsibility of directing the committee. An overview of the Army's needs showed eight air-cooled diesel engines of various sizes:

–	30 hp	Pkw Volkswagen (Kdf Wagen)
–	70-80 hp	Lkw 1.5 t, Lkw 3 t, Radschlepper Ost, Raupenschlepper Ost, kleines Kettenrad
–	110-120 hp	Lkw 4.5 t, leichter Panzerspähwagen, Zugkraftwagen 1 and 3 t
–	220-240 hp	Lkw 6.5 t, schwerer Panzerspähwagen, Zugkraftwagen 8 t, Panzerkampfwagen
–	280-320 hp	Zugkraftwagen 12 and 18 t
–	530-570 hp	Panzerkampfwagen
–	700-750 hp	Panzerkampfwagen
–	1000-1200 hp	Panzerkampfwagen

It was determined that standard cylinder sizes would be used. The indicated performance figures would be attained by utilizing a multi-cylinder arrangement (from 4 to 18 cylinders). It was originally planned to select two cylinder sizes of 1.1 liters and 2.2 liters displacement. In the end, however, three cylinder sizes were chosen: 0.8 liter, 1.25 liters and 2.3 liters displacement. It was expected that the cylinder of:

— 0.8 liter displacement at n = 2800 1/min would perform at 13 metric hp

— 1.25 liters displacement at n = 2400 1/min would perform at 20 hp

— 2.3 liters displacement at n = 2200 1/min would perform at 30-34 hp, with a supercharger at 40-45 hp.

Trials of a single cylinder test engine were to begin on 15 October 1942. Beginning in March 1943 4 cylinder engines in the 1.25 liter class — manufactured by Klöckner-Humboldt-Deutz — were subject to stringent tests at the Verskraft in Kummersdorf-Schiessplatz. Beginning in the autumn of 1943 4 and 6 cylinder engines were undergoing driving trials there in transport vehicles. However, it was still a long road to a multi-cylinder engine for armored vehicles. In actuality there was still none of this type of engine available for testing in an armored vehicle, making mass production as yet unthinkable. On 16 July 1942 Henschel agreed to the following level of production: December 1942 1, January 1943 5, February 10, March 15 and April 20. In order to avoid the continuing pressure of the Speer ministry Henschel would attempt to manufacture 66 Panthers by April 1943.

In the meeting of the Panzerkommission on 14 July 1942 it was agreed to install 100 clutch brake steering units in the Panthers on the condition that by the end of April 1943 all Panthers then produced would be reequipped with the single-radius epicyclic steering units.

By the 13th of July 1942 approximately 70% of the necessary parts for 1000 vehicles had been ordered. Bottlenecks were expected primarily in obtaining the road wheel arms, which were temporarily being manufactured by Siepmannwerke, Belecke/Möhne. The selection of the steering brake was not expected before 27 July 1942. A contract was awarded to Adlerwerke, Frankfurt am Main for 50 OLVAR drives developed by Maybach to be installed in the Panther. These were to be a test series. Thus equipped this version of the Panther tank would be designated Model B. However, these were not installed in the series production models. In 1942 ZF was able to deliver 61 AK 7-200 gearboxes for mass production.

The ten 8-point drilling machines for hull construction were built by the companies of Herkules, Siegen and Berninghaus, Velbert. The first hull drill press was delivered on 15 September 1942.

On 13 July 1942 it was determined that the single-radius epicyclic steering was to be THE steering mechanism for series Panthers. Reichsminister Speer had specifically directed that this drive was to be installed in the initial production run. The manufacture of 60 clutch-and-brake units was therefore halted on 14 July 1942. MAN hoped to conclude trials on the single-radius epicyclic steering by mid-October. This was simplified by limiting testing in high gear to two turning radii (80 or 115 m) instead of the three radii (50, 80 or 115 m) which would have provided the most accurate basis for a decision. It was therefore planned that two interchangeable gear sets be made for the first twenty to thirty single-radius steering units. In the final analysis it was determined that an 80 m radius in high gear would be the criterion. Around the end of July 1942 there was discussion of making use of the Reichsbahn repair facility in Berlin-Falkensee for Panther production. A decision had not yet been reached on whether to turn overall project planning over to Henschel or the Altmärkische Kettenwerk GmbH (Alkett).

On 4 August 1942 Maschinenfabrik Augsburg-Nürnberg made the announcement that it would begin construction of the first steel hull and requested that the other manufacturers of the Panther send their foremen and chief operators to Nuremberg for familiarization with the project.

Henschel reported on August 7th 1942 that it would assume production of the single-radius epicyclic steering unit nr. 48367. Construction would begin on 10 October 1942. It was expected that by 1 January 1943 the monthly output would be 110 sets.

The Rheinmetall turret for the second test Panther was not available in time, the turret housing only being finished on September 16th 1942. Final work on the turret

First MAN VK 3002 prototype during factory trials.

was completed by Rheinmetall-Düsseldorf.

On 18 September the announcement was made that 15 Panthers would be placed in a new, special category to ensure their timely completion (Stage DE (Dringliche Entwicklung or Urgent Development)). At the end of September 1942 Panther development ceased in the Daimler-Benz construction bureau at Berlin-Marienfelde.

The first mild steel hulled Panther chassis (prototype VK 3002) was delivered by MAN at the end of September 1942. It differed from the later mass-produced tanks in that the Panther V1 had the shock absorbers mounted on the first and eighth road wheel arms.

The 11th meeting of the Panzerkommission took place on 2 and 3 November 1942 in Eisenach. On the local training field of Panzer Regiment 2 Eisenach in Berka, an outpost of the Verskraft Kummersdorf for rough terrain, the following vehicles were to be demonstrated:

Two Henschel Tigers (one a standard production model, the other equipped with an electric ZF gearbox, type 12E-170, two Porsche Tigers, a MAN Panther (with turret), a Daimler-Benz Panther, a VK 3601 (Henschel), a Panzerkampfwagen III (ZW 40) with OLVAR gearbox, a Zugkraftwagen 1 ton class, a Zugkraftwagen 3 ton class, a Raupenschlepper Ost, two Radschlepper Ost, a Latil tractor, two Maultier 2 ton class (Bauart SS), a Maultier 2 ton class (Bauart Opel), a Maultier (Bauart Luftwaffe) an a 3 ton Lkw Opel A-Type. Also seen on the field were: a Panzerkampfwagen II (LaS 138) with electric ZF gearbox), a Panzerkampfwagen III with wide (eastern front) tracks, a Panzerkampfwagen III with rubber-conserving steel road wheels, two Panzerkampfwagen III with flame-throwing equipment and two Panzerspähwagen equipped for flamethrowing. A T-34 and a KV 1 captured from the enemy's forces were also on the field.

Speer drove the MAN Panther cross-country for more than 1 1/2 hours and had nothing but praise for its handling capabilities, even though at this time the vehicle was still equipped with the clutch-and-brake steering. The second MAN Panther, which was temporarily detached to the demonstrations, was equipped with a controlled differential discontinuous regenerative steering unit. These trials revealed that the differential operation was adequate even in rough terrain, without having to apply a steering brake while maneuvering.

On November 28th 1942 Henschel began production of the single-radius epicyclic steering units, starting with two pieces. After delivery to MAN the first of these was cleared for installation in the initial test vehicle on 4 December 1942.

The Maschinenfabrik Augsburg-Nürnberg, in November of 1942, was not in a position to deliver its Panthers as agreed upon. The first four production vehicles would be delivered in December. MAN planned to install a clutch-and-brake steering unit in the first 30 tanks. In December of 1942 Henschel was contracted for over 1100 single-radius epicyclic steering units. Maybach built 20 HL 210 engines for the pre-series Panthers.

A conference was held in the Armaments Ministry in Berlin on 17 December 1942, during which the special urgency of the entire tank program was emphasized. Regarding the Panther, Oberst Thomale, OKH/In 6 (Inspektion der Panzertruppen), determined that the mounting of the machine guns was unsatisfactory. Feasible solutions would also have to be found for the disposable fuel containers mounted on the rear hull, the winterizing equipment and the submersible qualities.

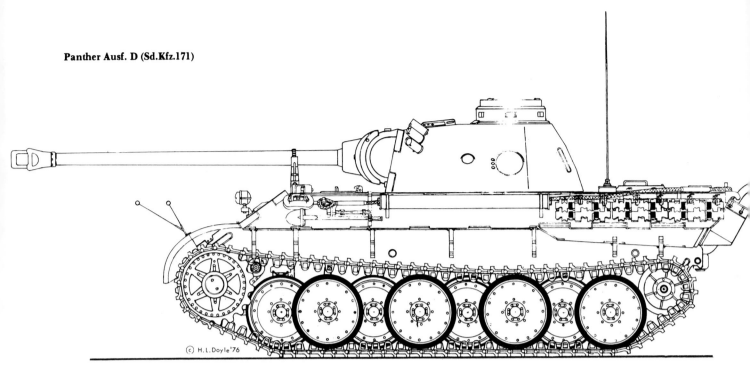

Panther Ausf. D (Sd.Kfz.171)

Thomale stated that he would be willing to accept the first 50 Panthers without amphibious gear if this equipment were delivered at a later date. Chief Director Saur of the Reichsministerium für Rüstung und Kriegsproduktion indicated that two of the initial Panthers were to be equipped with armor foreplating of 30 mm and 50 mm respectively. While it appeared that deliveries of the hull were secured, there were problems with turret deliveries. MAN received the first turret for the Panther to be delivered in November on 15 December 1942. The construction of the optics was only started on December 24th, 1942. Maybach announced that the deliveries of the engines were assured. The scheduled deliveries of the steering mechanism were tentatively seen as attainable. Saur made the determination that only the ZF synchromesh gearbox would be considered. Thomale indicated that the troops felt the spring suspension was too weak and the acquisition of strengthened road wheel arms was urgently pressed. It was anticipated that there would be a bottleneck in the manufacture of the road wheel arms. For the shipment of the fan drive by Fross-Büssing it was expected that a firm delivery agreement would be reached

The MAN pre-production models were recognizable due to the commander's cupola cut into the turret side wall.

This drawing from a propaganda magazine shows the crew disposition in the Panther. The only member not visible is the radio operator.

Postmark of the Alkett firm in Berlin-Tegel, which was responsible for numerous tank prototype developments. The company was not well known publicly.

on the part of MAN.

Saur expressly indicated that the construction of the Panzer III would not take a back seat to the Panther and that the planned production levels would be maintained at all costs. At this time all companies participating in the Panther program were also involved in mass production of the Panzer III.

Henschel had completed tooling of the first three Panther hulls on 4 January 1943 and these were made available for assembly. Important pieces were still missing for continuing work on the tank, such as the hand and foot controllers. As the supplier, MAN was informed by Henschel that the production had come to a standstill and could not be continued due to lack of supply of certain parts. The comprehensive changes to the single-radius epicyclic steering which were announced on 31 December 1942 caused a delay of two weeks. However, the risks of installing an untested steering mechanism design had already been taken into account.

On 1 January 1943 the Reichsminister für Rüstung und Kriegsproduktion made the announcement that various difficulties had been encountered by MAN and Daimler-Benz during the construction of the first Panthers. Henschel and Maschinenfabrik Niedersachsen-Hannover were directed to temporarily detach an expert to the two previously named firms in order to become acquainted with the problems of the Panther production.

At this time the planned manufacture of the Leopard reconnaissance vehicle was dropped in favor of the Panther. Hitler demanded research into the possibility of arming the Panther with a 7.5 mm gun of a length of L/100 or larger. In order to attain a higher penetrating force, he felt that correspondingly more powerful types of propellant charges should by used.

According to the Führer's directive of 8 January 1943, the Panther was to be built in Italy for the Italian forces in accordance with a mutual armament assistance program. Neither the Italian State nor the as yet undetermined license manufacturer would be required to pay any type of licensing fees. The drawings and construction data were given to the Royal Italian War Ministry.

Despite its best efforts, MAN was not able to deliver the promised four Panthers by the end of December 1942. Also, Rheinmetall had not as yet been able to tool up the turrets. New problems cropped up repeatedly with the chassis. The gears in the final drive were particularly troublesome. Here ZF, in order to achieve the strongest possible tooth root, had made use of a corrected tooth design — a design which showed a tendency for fatigue cracks even after only short distances.

P32:

The first Panther was delivered by MAN on 11 January 1943, operable and with radio equipment installed. By the end of January 1943 MAN was scheduled to manufacture 14 Panthers.

Henschel's schedule for the construction of its first Panther anticipated delivery for road testing on 18 January 1943. The first vehicles were delivered minus their winterizing gear and amphibious equipment. The Wegmann company stated that the first two Panther turrets it was constructing would be delivered on 15 and 17 January 1943. During January 1943 MAN conducted intensive testing of the single-radius epicyclic steering. Six tanks were on Henschel's assembly line at the end of January 1943. Daimler-Benz announced on 25 January 1943 that the accelerated transfer of ship engine manufacturing facilities from the Berlin-Marienfelde plant would be undertaken in favor of Panther production. A disposition filed by the Hauptausschuss Waffen (Main Weapons Committee) on January 29th, 1943 provides an interesting view of the tank production program outlined by the Reichsminister für Rüstung und Kriegsproduktion. Beginning in October 1943 it was expected that the following quantities would be delivered monthly:

— 150 light Panthers (with 50mm KwK 39 L/60)
— 900 heavy Panthers (with 75mm KwK 42 L/70)
— 75 Tigers
— 10 Mäuschen

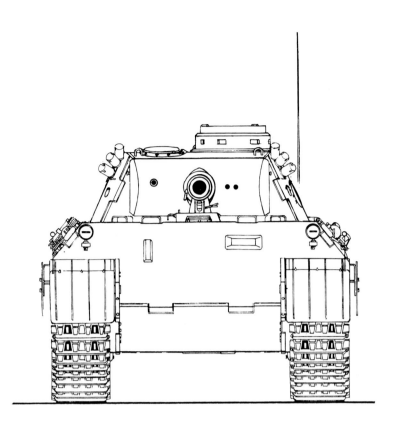
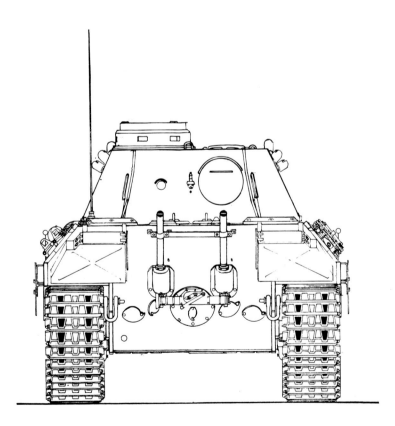

Four-view of the initial production model of the Panther tank, Ausf. D (Sd.Kfz.171)

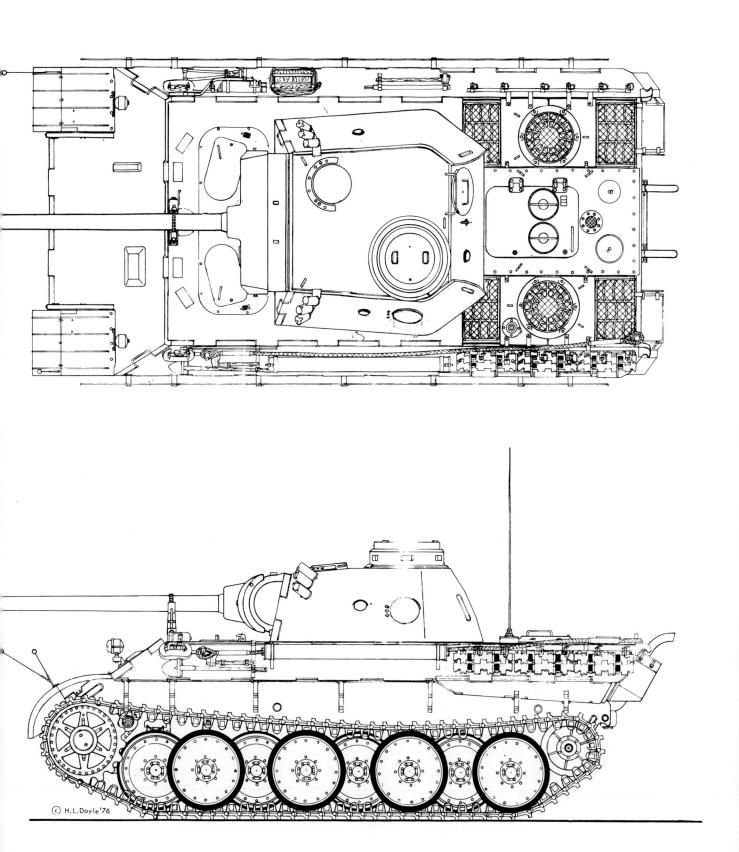

Panther tank with 75mm KwK L/100 (proposal)

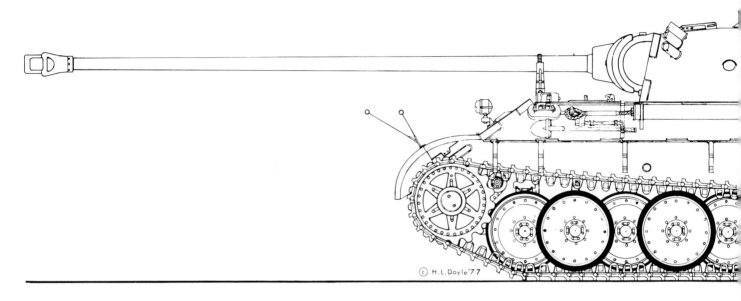

— 250 light Sturmgeschütze (on Panzer III chassis)
— 250 heavy Sturmgeschütze (on Panther chassis)
— 200 light self-propelled guns (on Panzer II chassis)
— 150 heavy self-propelled guns (on Panzer III/IV chassis)
— 15 ultra-heavy self-propelled guns (on Tiger chassis)

The first and second MAN mass-produced Panthers were delivered to Panzer-Abteilung 51 at the Grafenwöhr training area on 24 January 1943; the third arrived on the 26th of that month. Since the first two vehicles were to be turned over to the troops for training, only the third was available for industrial trials. The fourth MAN Panther was delivered to the Verskraft in Kummersdorf for the initiation of testing, which had not yet begun. The vehicles had not been officially appropriated, but were handed over to the troops on the personal orders of Speer. Official acceptance would take place later at the Grafenwöhr training area. The right and left underside corners of the turret housing would have to be cut back by 30 mm, since these corners struck the closed driver and radio operator hatches. Furthermore, problems were also encountered with the traverse mechanism in the turret, which could not rotate when the Panther was at an angle.

All vehicles being produced by the four manufacturers carried the designation Panzerkampfwagen Panther, Ausführung D (Sd.Kfz.171) (on initial drawings it was still shown as Ausführung A). 850 Panther Ausführung D models were built.

The first Panther hulls on MAN's hull drilling mill.

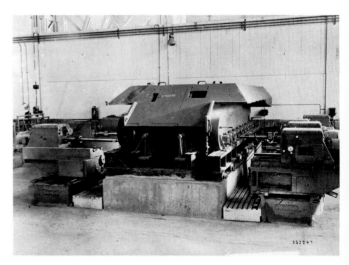

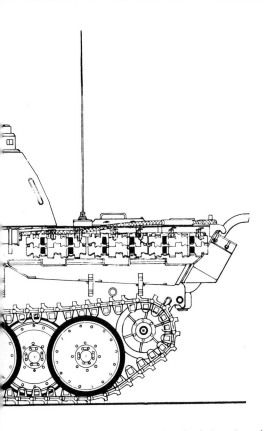

Driver's seat in the Panther.

The driver's and radioman's seats separated by the transmission and steering differential. In the first Panther model, the radio operator only had a machine gun flap.

The Panther tank consisted of the chassis and the 360 degree traversable turret. In the hull, a firewall separated the fighting compartment from the engine compartment. The hull contained the engine and the drive units, as well as the running gear mounts. Various hatches, louvers, ports and other openings were to be found on the hull and turret. The resulting combat weight of 45 metric tons — originally planned at 35 tons (!) — and an engine performance of 700 metric hp gave the vehicle a power to weight ratio of 15.5 hp/ton.

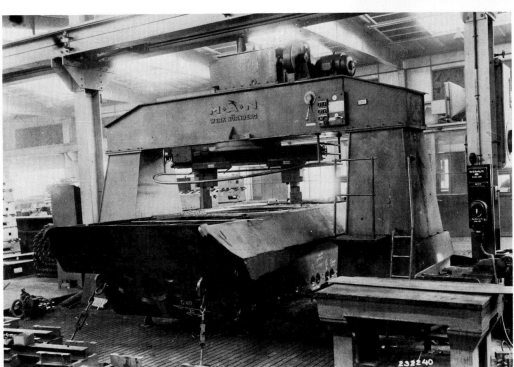

Here the seat of the turret race plate is being machined.

By the beginning of 1943 the Panther was in mass production with four manufacturers.

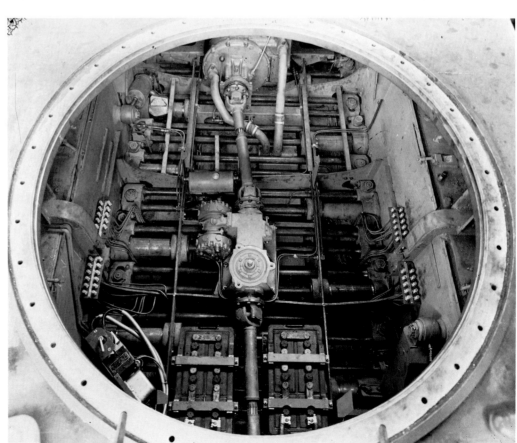

A view into the hull shows the double arrangement of the torsion bars, as well as the drive shaft from the engine to the transmission.

Details of the Panther production process at the Maschinenfabrik Augsburg-Nürnberg in Nuremberg.

The finished chassis is lowered onto the tracks.

One of the first Panthers built by MAN shows the layout of tools and other equipment on both sides of the vehicle.

The engine, a liquid-cooled 12 cylinder V-type combustion engine type HL 230 P 30 was a development of the Maybach-Motoren-Gesellschaft in Friedrichshafen. The concept of the HL 230 was an attempt by Maybach to meet all the requirements necessitated by a combat tank in the 30-40 ton class. The demand for an engine of smaller dimensions was certainly met. The engine was hardly longer than the proven 12 cylinder HL 120 engine, already in operation with the Panzer III and IV. The unusually short length contributed to very compact dimensions of the Panther's engine compartment and favorably influenced both the overall length and the weight. The fighting compartment, therefore, constituted 63.7 per cent of the total interior space. The engine weight was also kept low, even though the original plans calling for an alloy HL 230 engine in production models would have to be changed to a cast iron design. The cast iron block was characterized by its rigidity. Engine performance was 1.72 kg/metric hp with a displacement of 30.5 metric hp/liter and an average piston speed of 14.5 meters per second.

The exceptionally long development time normally expected of such a lofty undertaking was not available, since the Panther would have to be developed and go into production quite rapidly. Therefore it was not possible to conclude development and undertake the necessary testing of the vehicle.

The troops had come to trust Maybach engines, since the various motors produced for the newly-developed tanks had exhibited few notable difficulties. Given the rapid expansion of the Wehrmacht, it was extremely rare to find a tank type which had passed through all its developmental stages prior to reaching the front lines. As opposed to these engines, the current series was plagued with all kinds of problems which would only be rectified later in the manufacturing process. The bottom end bearings, which on the HL 120 series were arranged in a row on the crankshaft, had been replaced by a forked connecting rod and a solid end connecting rod — both on the same crankshaft pin; this reduced the overall size of the engine by half a cylinder diameter. The initial difficulties caused by such an arrangement could be corrected. However, it was not so easy wrestling with the problems of

Side view of a Panther Ausf. D with the initial arrangement of the commander's cupola. Machine pistol plugs and communications port can be seen in the side wall of the turret.

A Henschel-produced vehicle prior to laying on the tracks.

Panther tank, Ausführung D.

Both Tiger and Panther tanks were built simultaneously at Henschel. The Panther already sports the Sandgelb basic coat introduced in 1943, whereas the Tiger still wears the older Dunkelgrau.

The first vehicle delivered to the Henschel testing grounds, which was prepared for submersion trials in Haustenbeck.

At this spot the track struck against the tongue mounts of the side skirts and tore them away.

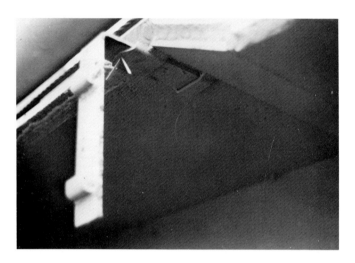

The mountings for the side track skirts proved to be too weak.

The track skirts were pushed into each other and twisted.

the distance between cylinders being so significantly reduced in comparison with previous engine designs. All these teething troubles, which could have led to major problems in operations with the front-line troops, were eventually corrected by the end of the Panther's development cycle to the point where the HL 230 enjoyed nearly the same level of reliability as Maybach's other engines. The engine's cylinders were arranged in a V-form and their lubricated cylinder sleeves were interchangeable. The crank assembly consisted of the crankshaft and flywheel, the piston rods and the pistons; these were formed of metal alloy. The crankshaft rolled on seven sets of bearing races. In later versions the crankshaft was supported by an eighth set of bearings behind the vibration damper. A forked connecting rod was affixed to each crankshaft pin, which also supported a link rod for the opposite side of the cylinder. The overhead camshaft ran through the length of the cylinder head cover and operated the intake and exhaust valves by means of rocker arms. In order to facilitate better cooling, the smaller exhaust valves were hollow and filled with a special type of salt.

As with all German tanks, a dry sump lubrication system was used to ensure trouble-free lubrication of the engine parts when the tank was operating at a steep angle — and in view of the imposed low profile design of the Panther.

The equipment used for regulating the temperature of the engine coolant and the transmission fluid could be shut off for underwater travel. The radiators, the fans and the air passages were separated from the engine compartment and watertight. Two vertically positioned radiators and one blade fan between the radiators were installed on both the right and left side of the engine, between the engine and the hull walls. When the Panther was travelling submersed, the fans were shut off and the space was flooded. The fans were driven by the engine via a drive having two gears for both winter and tropical operations.

A torn out mount for the track skirts.

The hatch in the middle of the rear hull wall seen here could not be sealed properly initially, and it was necessary to develop a new seal design.

The container for the gun barrel cleaning equipment, seen fixed above the side skirt, proved to be inadequately sealed during amphibious testing.

Residual grit or sand made it difficult to close the driver's flap.

The torque was transferred from the fan hub to the fan wheel through a slip clutch, which prevented the power from engine acceleration from being directly relayed to the fan drive and thereby minimizing the potential for damage to the drive. The fluid level of the cooling system was maintained by the use of an overflow/expansion tank in the rear of the vehicle. The coolant was poured into this expansion tank.

The wet air filter, either with its lubricated metal trap weave or filled with oil to filter the engine combustion air, was originally installed in German tanks and required an inordinate amount of maintenance. This filter type led to many engine breakdowns during tank operations in North Africa and Russia. The method of cyclone dust removal, commonly used in the gas refining industry, was explored as a possible solution for removing dust from the engine combustion air of armored vehicles. Professor Dr.-Ing. Feifel, of Vienna's Technische Hochschule, was at the forefront in exploring the theory of cyclone dust separation and had formulated mathematical calculations for determining the most ideal path of circulation flow in the cyclone chamber.

The result of this developmental work was Professor Dr.-Ing. Feifel's cyclone filter and possessed outstanding qualities; with the Feifel filter it was possible to methodically regulate the process in the vortex core and base in order to achieve optimum dust separation in the tank's engine.

The Filterwerk Mann & Hummel GmbH in Ludwigsburg assumed mass production of this filter, which was installed in the Tiger and Panther tanks. The separator chamber in the cyclone filter contained no moving parts. The airflow resistance was independent of operating time; at maximum r.p.m. the resistance was approximately 150 to 200 mm WS and dropped off with reduced r.p.m without any appreciable loss in the quality of dust separation. At maximum r.p.m. the filter had a 99% dust separation effectiveness. The Feifel filters were used

Frontal view of the Panther equipped for fording trials.

Original form of the gun mantelet, with barrel travel lock.

Stowage boxes were mounted on both sides of the rear hull for transporting the crew's baggage. They, too, were prone to leaks. Panther, ready for delivery.

Panther Ausf. D tanks during delivery.

For the first time the Panther appeared in the recognition leaflets. The vehicle below still has the ball-shaped muzzle brake.

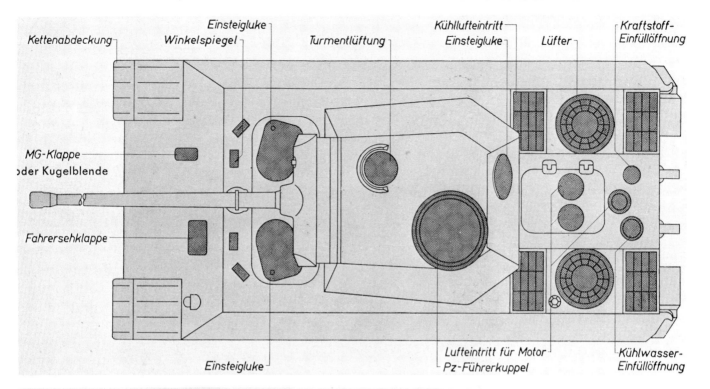
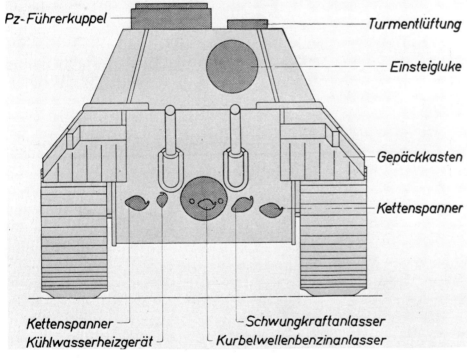
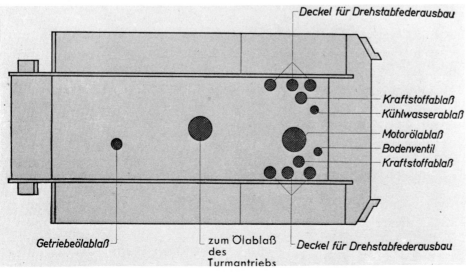

A Panther schematic shows the overhead view, the rear view and the tank hull from the bottom. Entry and maintenance hatches are indicated in dark gray.

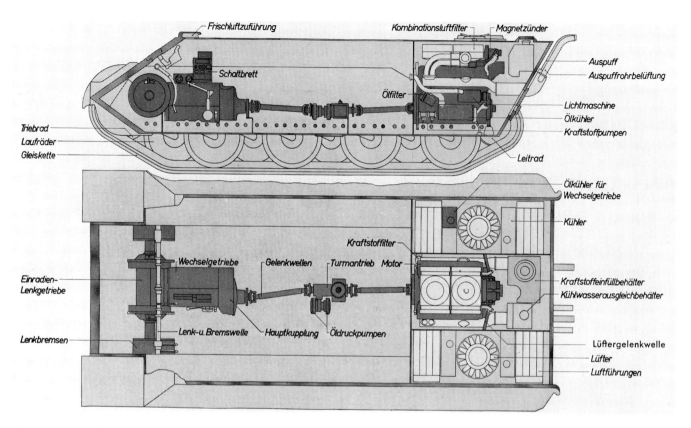

The power train of the Panther in both schematic and perspective views.

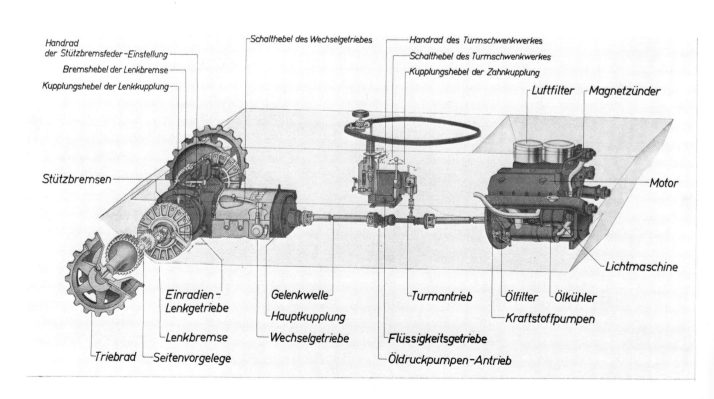

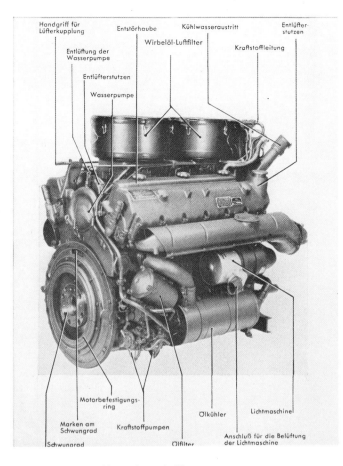

Engine, generator side (cyclone air filter)

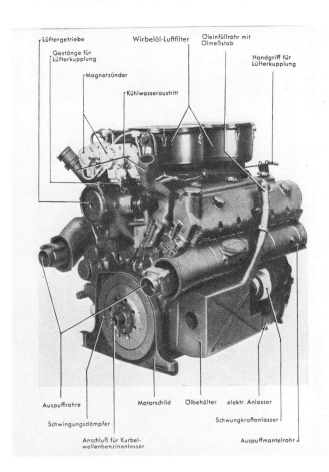

Engine, starter side (cyclone air filter)

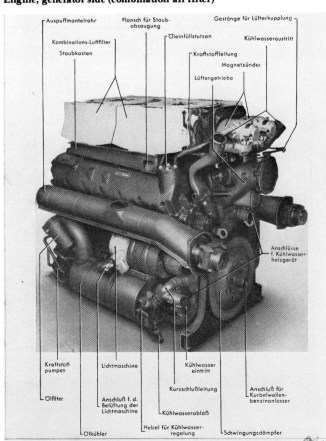

Engine, generator side (combination air filter)

Engine, starter side (combination air filter)

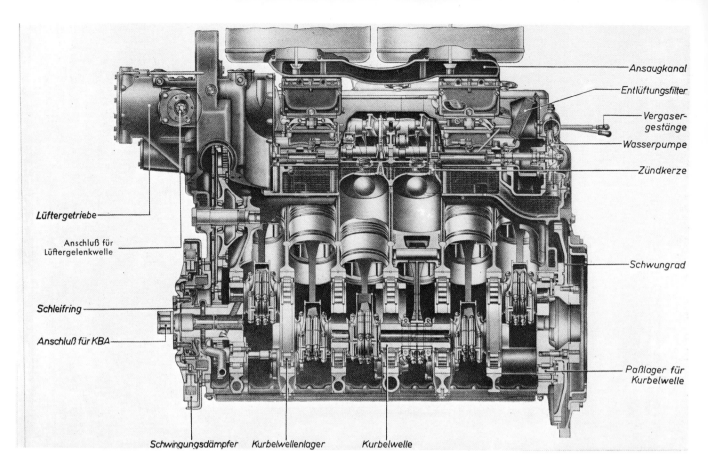

Above: engine, sectional view.

Below: engine, cross-sectional view.

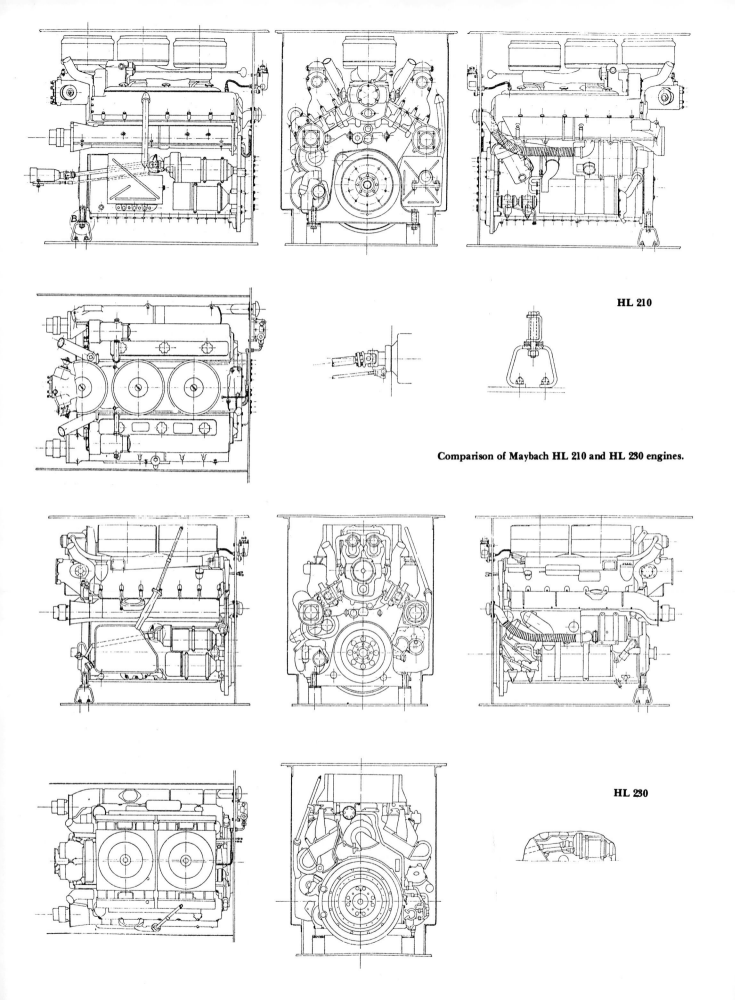

HL 210

Comparison of Maybach HL 210 and HL 230 engines.

HL 230

Maybach drawings in 1:2 scale showing the cylinder arrangement of the HL 230 engine. The narrow spacing between the cylinders was a result of the use of a disk-type crankshaft. Not once did the small water spacing lead to piston damage; it was not necessary to take special measures.

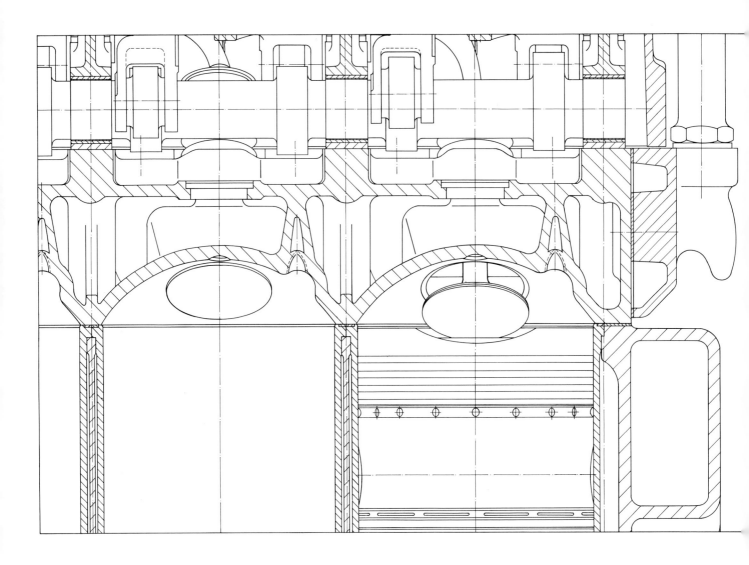

On the HL 230 engine the cylinder head's crankshaft housing was sealed against oil and water by a Reinzolit gasket (1). The cylinder sleeves (2) butting up against the cylinder head — i.e. the compression chamber — were sealed by rectangular soft copper rings (3 and 4). These rings were forced down into the vertical section of the upper part of the cylinder sleeve as the cylinder head was tightened down. Sketch 4 shows two of the rings fixed in the vertical section.

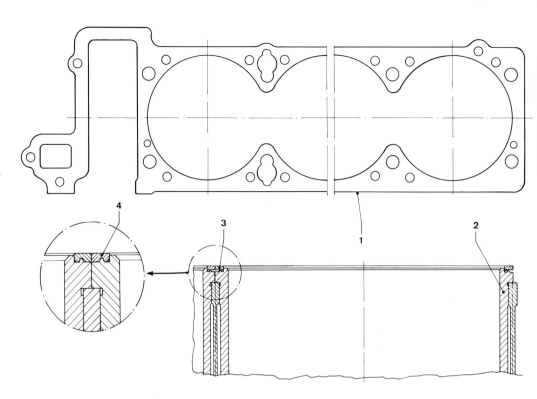

Comparison of the Maybach HL 230 P 30 (Panther) and the HL 230 P 45 (Tiger).

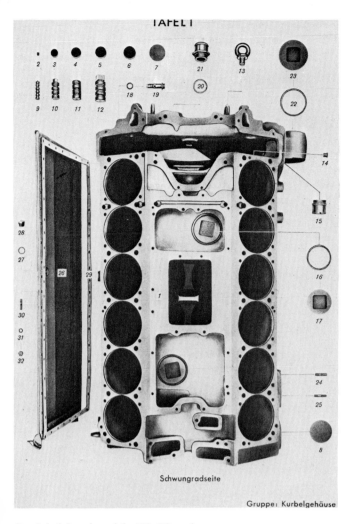

Crankshaft housing of the HL 230 engine.

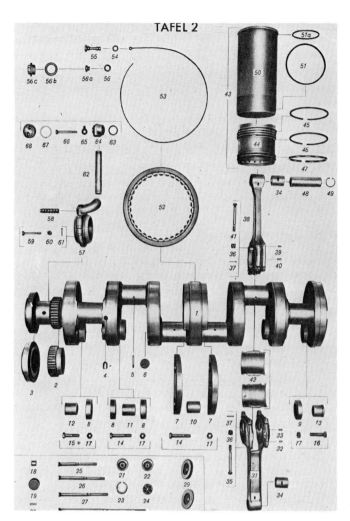

Details of the HL 230 engine

Overview and details of a disk crankshaft.

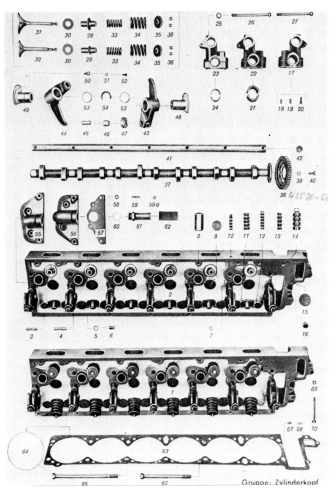
HL 230 engine — cylinder head

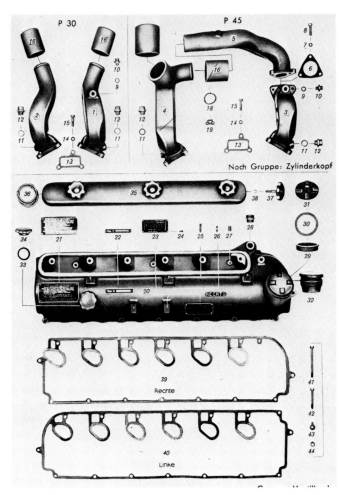
HL 230 engine — valve cover

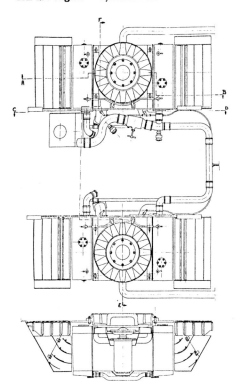

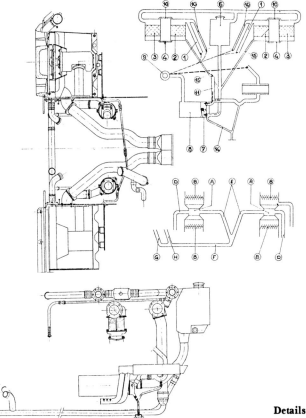
Details of the cooling system.

50

exclusively as a pre-filters, the "cyclones" being operated by means of a parallel connection. These filters were installed on the series models of the Maybach HL 230 P 30 engines. The dust was automatically drawn out of the settling space by the fans of the cooling system. In doing so, a partial maintenance-free operation was achieved (the primary oil-bath filter, used for improving dust filtration even further, still required servicing). Together with the firm of J. Hückels and Sons of Neutitschein, Dr.Ing.h.c. Ledwinka of the Tatra company pushed for thorough studies into replacing the oil-bath filter in the Panther with a maintenance-free felt filter. However these studies were never concluded.

The exhaust manifolds became extremely hot during operations and were therefore covered with cooling sleeves or jackets. The cool air was drawn through the space between the manifold and the jacket by means of a pipe running from the fan. In order to make this cooling arrangement more effective the air was forced through the exhaust jacket pipe on the left and flowed into the open air through two pipes mounted on either side of the left exhaust pipe.

The five internal fuel tanks held roughly 730 liters (560 kg), of which 130 liters were considered reserve. This total amount gave the vehicle a highway range of 250 to 270 km at 23 to 26 km/h; the reserve gave 40 to 50 km. Two double head fuel pumps forced the fuel to the four double-downdraft carburetors adapted for cross-country operations.

Engine and gearbox were connected via two cardan shafts, separated by the turret power traverse drive. The turret drive shaft lay between the cardan shafts in a housing, fromwhich the turret drive was powered along with the two hydraulic pumps for operating the steering brakes. The turret traverse was driven from the first set of bevel gears via a short drive shaft and a hydraulic drive (Böhringer-Sturm drive).

Since the Panther was designed to be amphibious, the engine compartment was made watertight — sealed off from the radiator and fan compartments as well as separated from the fuel tanks. The engine compartment was inadequately ventilated, bathed in its own combustion air in addition to the already warmed cooling air passing

Cyclone combustion air filter.

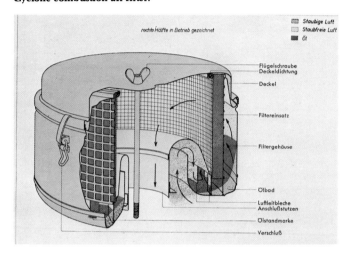

Combination combustion air filter.

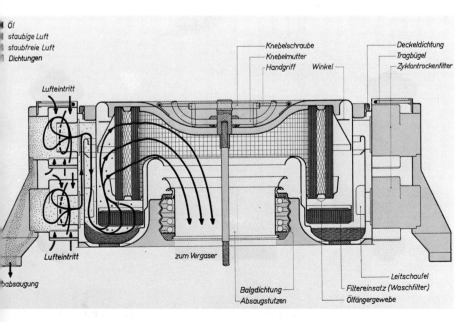

Cyclone filter battery.

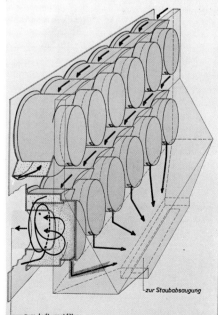

through the cooling sleeves of the exhaust manifolds. This arrangement led to many engine fires caused by non-insulated fuel connectors, particularly in the first models. The situation was partially remedied by drawing off the highly flammable bottom layer of gasses by using lower pressure air in the suction valve chamber beneath the side-mounted fans. Even this solution did not prove completely satisfactory initially.

The main clutch in the form of a triple plate dry clutch separated the power transfer between the engine and the gearbox.

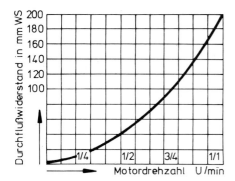

Der Durchflußwiderstand des Zyklon-Filters ist niedrig.

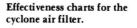

Effectiveness charts for the cyclone air filter.

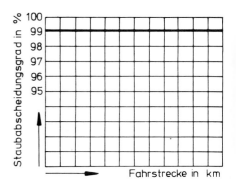

Der Ausscheidungsgrad des Zyklon-Filters ist bei gleichbleibender Belastung von der Betriebsdauer unabhängig und bleibt stets gleich gut.

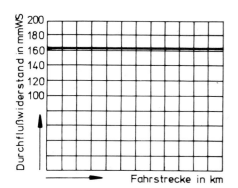

Der Durchflußwiderstand des Zyklon-Filters ist bei gleichbleibender Belastung von der Beanspruchungsdauer unabhängig.

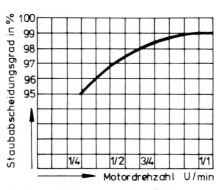

Der Ausscheidungsgrad des Zyklon-Filters ist von der Drehzahl des Motors weitgehend unabhängig.

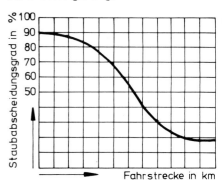

Der Staubausscheidungsgrad eines gewöhnlichen Naßluftfilters sinkt nach kurzer Betriebsdauer stark.

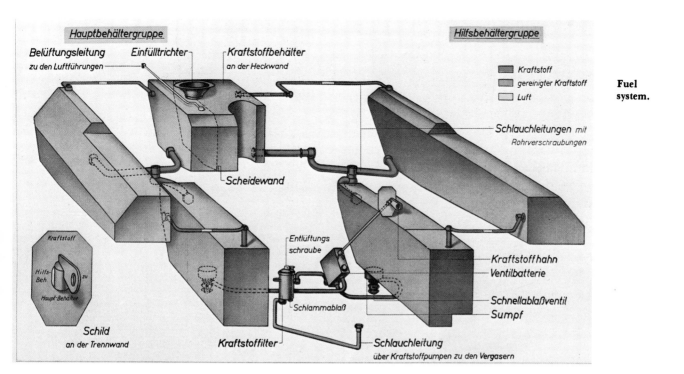

Fuel system.

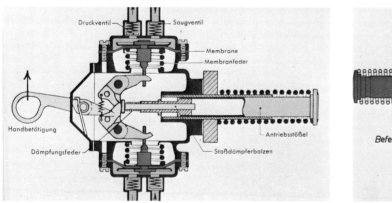

Operating characteristics of the fuel pump.

Fuel pump.

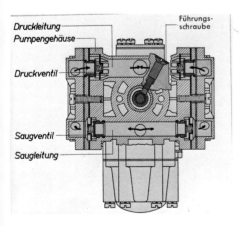
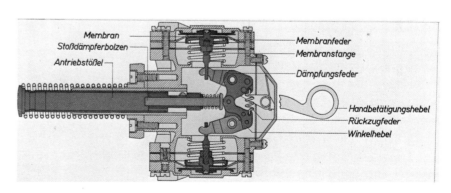

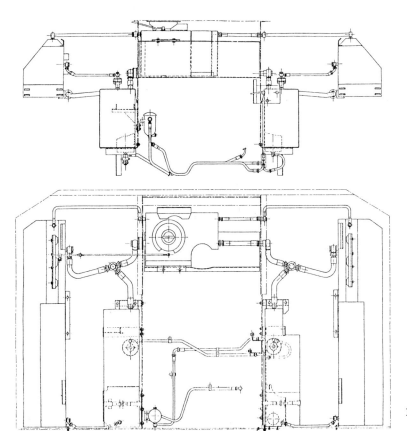
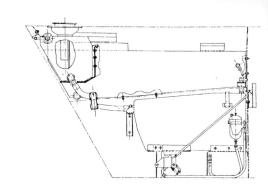

Fuel system as installed in the vehicle.

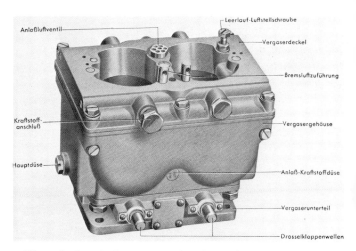

All-terrain downdraft carburetor, side view (Solex)

Carburetor, top view minus cover.

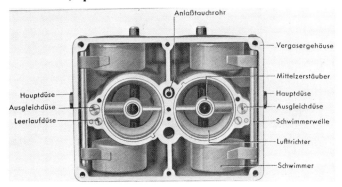

Both the Maybach-VARIOREX and OLVAR gearboxes were very costly in terms of material quality, ball bearings, refinement in construction, construction time involved and maintenance. Other than these two designs, there were no semiautomatic or fully automatic shift gearboxes being manufactured in the required size, nor was there any suitable transmission which did not interrupt the drive power. The familiar multi-plate disk clutch and hand-operated gear shift were therefore selected. The Zahnradfabrik Friedrichshafen pushed for the design of a simplified gearbox, primarily with regards to maintenance. Graf von Soden, the well-known transmission expert, penned his thoughts on tank transmissions in a 1942 memorandum and called for the "absolute most simple construction."

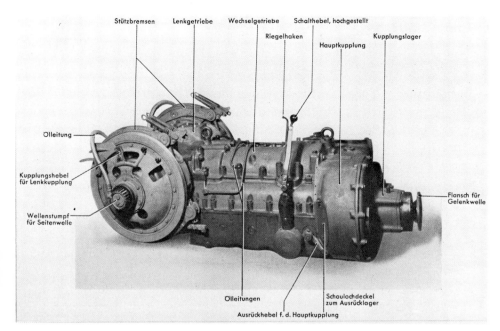

Transmission, left side.

There were two paths being pursued in the Oberkommando des Heeres (Army High Command) regarding transmissions; the one pursued by the Heereswaffenamt emphasized automatizing the shifting process, while the Inspektion der Panzertruppen (In 6) stressed a simple manual transmission with few gears. Discussions arose as to whether it would be feasible to equip a 45 metric ton vehicle with a manual transmission which wouldn't tax the driver beyond his physical limitations. The Zahnradfabrik Friedrichshafen began work on a series of drafts; Graf von Soden and Dr.-Ing. e.h. Maier settled on the triple shaft design AK 7-200 transmission. The design of this seven-speed all synchro-mesh AK 7-200 tank transmission was carried out in an unusually short time span. In February of 1942 initial work was begun on the design; in August of the same year the first two transmissions stood awaiting delivery. These two test units initiated the immediate run of the first batch of 5000 transmissions, even though up to that point a manual transmission of 700 metric hp was virtually unheard of. The minor defects initially encountered were corrected over a short period of time. ZF met the demand for a production rate of 1000 units per month by constructing a new manufacturing plant in Passau. The new "Waldwerke GmbH" facility, generously supported financially by the OKH, began producing tank transmissions starting in mid-1943.

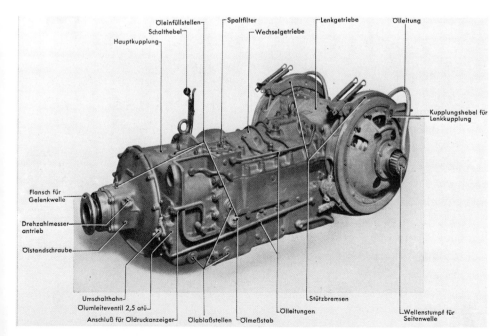

Transmission, right side.

Another branch site was established in Schlettstadt, Alsace. Up to this time, all spur gears which meshed were contained within the transmission case, a practical if costly solution. The normal alternative was to mount the spur gears on the shafts, as was done on the AK 7-200. This did not change the fact that such transmissions were not the ideal solution for these heavy vehicles with their high rolling resistance. The transmission was a seven speed all synchro-mesh design with synchronization for all gears excepting first gear and reverse.

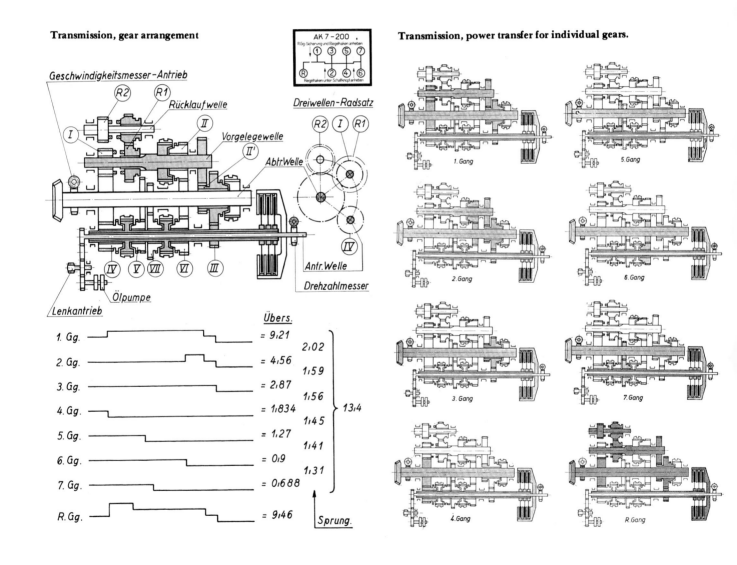

Since it was envisaged to produce the Panther in large numbers, production costs of the various subassemblies would have to be kept to a minimum. Instead of the double-radius epicyclic steering in the Tiger tank, it was considered that the single radius steering would be adequate; it had already proven effective in the Panzerkampfwagen 38(t) and permitted a large turn radius by using epicyclic gearing to reduce the speed on one side of the vehicle. Tighter radii were achieved by disengaging and applying the brake to the track on the inside of the curve. This type of steering was a vast improvement over a clutch-and-brake steering, even if it wasn't as appealing as the double-radius steering. If it had been possible to foresee what difficulties the final reduction gearing was to cause, it would have been a much better solution to have selected a more expensive final drive which provided a greater degree of reliability. In the end, the final drive proved to be too weak to handle braking with the Klaue disk break when steering through tight curves. The use of epicyclic gearing for the final drive hinged upon the bottleneck being encountered in the supply of gear cutting machines for producing the hollow gearing. When passing judgement on the double-spur final reduction gear it should be noted that the high-quality steel originally planned for the spur gears in the final drive was not available for mass production and was unexpectedly replaced by VMS 135 (today 37 MnSi5) tempered steel (not as suitable for this purpose). The steering unit itself

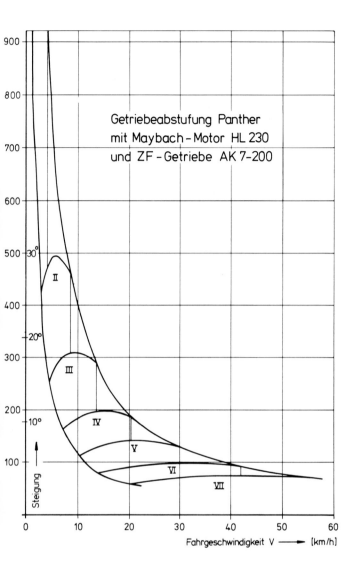

Gear progression of the Panther.

Switch system of the steering differential with final drive and their tooth numbers.

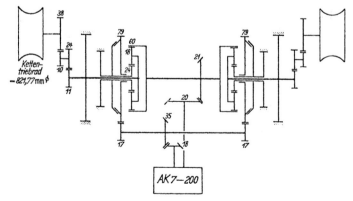

The noise of the forward gears was kept to a minimum by making use of angled-tooth gearing. The individual gear pairs were in a constant state of mesh in all gears. Synchronized shifting took place by means of sliding sleeves which were engaged with the shift lever. During shifting the cog wheels to be shifted were first brought to the same r.p.m. by a bevel gear pre-selector, only then was a set connection made between the drive shaft and the gear. During down-shifting the pre-selector was assisted by gas given at intervals to the engine in gliding downward; during up-shifting it was supported by double clutching. The transmission was designed for a top speed of 55 km/h and for 4.1 km/h in first gear. It had an final drive ratio of 13.4:1.

Production of steering differentials.

performed well with only minimal problems. It consisted of a primary bevel gear drive, two epicyclic gears, the bevel gear control drive, two spur gear pairs and two support brakes each having a control clutch. The center solid gears in the epicyclic gearing could be arrested by the brakes either individually or collectively. In order to facilitate steering through tight turns each track could be halted by a solid disk brake, once the support brake had been released and the steering clutch disengaged. Also, by releasing the support brake a single center gear could be set into motion from the control drive via a control clutch and a spur gear pair against the main drive's direction of rotation. The affected track would be slowed down and the tank would then make a single arc of a fixed radius for every gear engaged — hence the term "single-radius" steering. This type of steering required little strength and

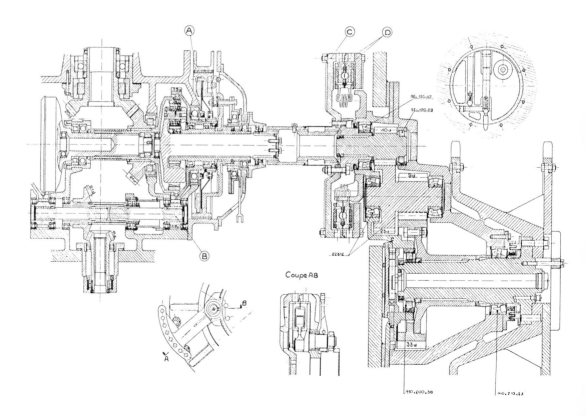

Steering differential, steering brake and final drive, with drive sprocket for the Panther tank.

could be used in all normal situations.

In order to keep the effort to a minimum during manual shifting, the operation of the support and drive brakes was enhanced by a hydraulic system. Tight turns required that the brake on the inside of the turn be applied in order to slow the track down or bring it to a stop, as needed.

Given the unfamiliarity with the speed/weight combination of the Panther tank, conventional brakes were simply not able to provide the necessary braking torque. The solid disk brake developed by Dr.-Ing. Klaue and manufactured by the Argus Werke furnished the required deceleration values in addition to demonstrating favorable endurance characteristics. The brake disk was designed as a U-shaped ring, heavily ribbed externally. Two fixed brake rings, kept apart by bearings in angled grooves, provided the braking power to the U-shaped brake ring. The large surface area of the brake rings meant that the wear of the brake lining (attached to the brake rings) was relatively low and allowed for the heat buildup to be satisfactorily bled off. The expansion force caused by the bearings in the sloped grooves provided the basis for the previously unheard of braking power.

Steering brake, components of the solid disk brake

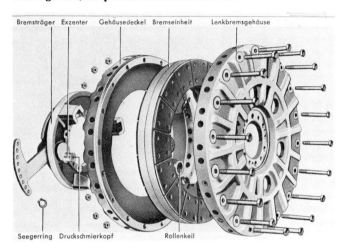

Final drive and drive sprocket.

Steering brake, sectional of the solid disk brake.

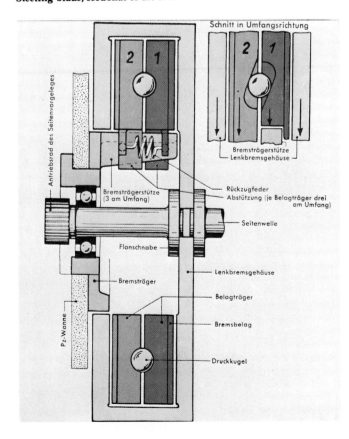

If the steering levers or brakes were handled in a rough manner such great momentum could build up as to cause breaks in the weak gear teeth or to the mountings of the under-sized final reduction drive shaft.

Steering brakes were not applied when making tight turns with a double-radius steering mechanism as mounted in the Tiger, thereby limiting the occurrence of such high momentum to the relatively rare times when emergency braking was necessary.

The heat buildup of the steering brakes was drawn off by a rectangular channel in the hull floor which ran beneath the engine to the right fan compartment.

The final drive served as a further reduction of the driving r.p.m. It contained two spur gear sets. The housing was sealed off by labyrinth packing at the point of the drive sprocket hub. The drive sprocket was joined to the final drive by its drive sprocket shaft. The toothed rims on the drive sprockets could be replaced without actually having to remove the sprockets themselves. The final drive (gear teeth and bearings) was the weakest part of the Panther. It was a risky proposition to use a spur gear system for transferring the drive power — especially considering that the available steel during the war did not have a particularly high stress tolerance. A better solution would have been to use an epicyclic gear system; a prototype final reduction drive using planetary gear reduction had already been tested and had performed flawlessly. However, as mentioned previously, a shortage of gear cutting machinery for the hollow gearing prevented this type of final drive from being mass produced. In order to bridge the gap a final reduction gear system was installed in front of the main gear drive, but due to installation restrictions its mountings were far too weak and could not be strengthened. Because of gear teeth being under too great a load and the weak mountings, the gears were pushed out of alignment — virtually guaranteeing mount and tooth breakage.

The general consensus of the industry was that inner-toothed gear wheels could not be produced due to a lack of proper machinery. This meant that a final drive using planetary gear reduction and pre-selector spur gearing — found to be reliable in company testing — could not be installed in production tanks. All attempts to improve the final drive met with failure, despite the offers of a special bonus as an incentive. The housings, which initially had proven too weak and whose outer mounts had been bent out of alignment by the track's pull, were eventually replaced by stronger ones.

What was the state of running gear development for German tanks prior to the introduction of the Panther? The Panzerkampfwagen III and IV originally were of limited usefulness to the German armored forces. The chassis of the Pzkpfw III was definitely an asset —after the double-bogie running gear of the initial series vehicles did not prove effective. The tank was relatively reliable with the exception of the rubber tires on the wheels (which had too small a diameter). Until the process of cloudburst hardening had been invented the torsion bars of the Pzkpfw simply did not have enough fatigue resistance. The vehicle was basically too small to accommodate a 75mm L/50 gun in its turret — meaning that when the Russian T-34 appeared the Pzkpfw III lost its effectiveness. In spite of a few flaws in mounting the gun, however, the chassis turned out to be acceptable as a Sturmgeschütz platform and was used with success in this role up until the end of the war.

On the other hand, the Pzkpfw IV chassis was large enough to carry a 75mm L/50 gun in its turret. However, the chassis revealed a number of basic technological design flaws. Once it was determined that the better Pzkpfw III chassis was too small for the 75mm L/50 gun, production of the Pzkpfw IV intensified. At this point it no longer became possible to alleviate the weaknesses of the Panzer IV's chassis. Practically speaking, it would have meant a new design of the steering brakes resulting in a total redesign of the steering unit, the running gear, the hull, the radiator and the engine compartment. This approach was repeatedly studied and discarded. The road wheels were of too small a diameter to provide the rubber tires with even a barely acceptable lifespan. The tracks were too narrow. The leaf spring bogie wheel suspension frame no longer mirrored the current state of technology. Tests using a volute suspension did not offer any useful results.

The band brakes used in the steering mechanism were technologically obsolete, less reliable and required a considerable amount of maintenance. The engine and transmission operated in a satisfactory manner. The combustion air filtration proved entirely inadequate in light of the radiator and cooling fan arrangement. The spatial separation of the fans and radiators automatically resulted in a poor effective cooling temperature.

The Panzer IV was clearly inferior to the Russian T-34 in all crucial areas. Given this situation it was imperative that a new, significantly more capable tank be developed

---

* As of this time, the ballisticians had not yet been able to assert themselves over the tacticians, who did not want to see the maneuverability of the tank be restricted in favor of the barrel overhang of the gun.

as rapidly as humanly possible.

This new model was designed in an inconceivably short amount of time in the form of the Panther tank. This chassis made great advances into the most diverse areas of transportation technology. Some of these ventures were a complete success; some suffered minor setbacks but were worked into acceptable solutions; some would only reach a mature stage of development in the closing days of the war, and some would only fulfill their purpose in very restricted conditions. The greatest accomplishment — albeit with reservations — was in the creation of the Panther running gear. This running gear established an entirely new standard for the performance of tracked running gear of the time and, in spite of its complicated design — both overall and in detail — seldom led to truly serious difficulties. It was the design of Professor Dr.-Ing Lehr*, a notable researcher in the area of oscillation studies and fatigue limits. Professor Lehr possessed the unique quality of not only having the highest scientific abilities but also having a clear vision of what was feasible in practice.

Professor Lehr operated on the principle that the suspension of a tank should permit the vehicle to safely, rapidly and with great maneuverability cross even a battlefield of extremely uneven terrain. From this thinking stemmed the question of how great the vertical displacement and maximum acceleration should be, given the varying terrain of a battlefield landscape. A long-range goal was established wherein it was determined to have a tank sprung in such a manner that it would be able to traverse at high speed an area having ground undulations up to a height of 500 mm, without the suspension breaking down. The suspension introduced with the Panther enabled the vehicle to overcome an RMS elevation (an indicator of surface roughness) of 150 mm. This objective was attainable only by taking advantage of all the options then available and by utilizing high-cost technology. A depth of 300 mm required a bump of 300 to 350 mm, calculated from the static laden position of the wheel to the point of its upper travel limit.** In order to study the problems of overcoming an RMS value of 300 to 350 mm, a test vehicle having a 400 mm stroke and developing 40 to 45 kph cross-country was turned over to the WaPrüf 6. Instead of the Panther's normal "hairpin" double torsion bars, three torsion bars were arranged in a row. The question of what speeds the Panther was expected to reach under combat conditions was clearly answered in message #32/43 g from the Generalinspekteur der Panzertruppen on 12 August 1943:

"Regarding all-steel resilient running gear: The Gen.Insp.d.Pz.Tr. requires that all parts of the running gear now being developed for armored tank vehicles (Tiger, Panther, Panzer IV and light Panzerjäger IV) meet the following criteria:
30 km/h cruising speed on the battlefield
40 km/h maximum speed on the battlefield
This demand, founded on the basic principles of tank warfare, will in no way be restricted by the generalization that in reality a tank's speed may actually be much slower on the battlefield."

Visibility could be improved significantly during high speeds by the use of gyro-stabilized optics for the tank crew which would keep the field of view level independent of the tank's vertical motion. Furthermore, it was expected that a tank having good suspension would provide a suitable observation platform while crossing rough terrain. In any case, the gunner should be able to maintain the target in his sights at all times. This would not be the case if he were constantly having to pull his head back to avoid being struck by the sight at every ripple in the terrain. A prerequisite for circumventing this drawback was ensuring that the vertical displacement remained less than approximately +/-2 degrees and that the force upon the gunner's head by roughly 40% lower than the vertical acceleration (b 0.4 g) even during the hardest jolts. This was not easy to accomplish, although the Panther went a long way to doing so by arranging the shock absorbers on the first and final swing arms.

---

* Dr.-Ing. habil. Ernst Lehr, born on 4 July 1896, killed in an air attack on Berlin on 25 March 1944.

** Testing was done at the Verskraft in Kummersdorf on a concrete track, on which undulations of 100 mm height at intervals of 6 meters were laid.

Spring layout in the Panther.

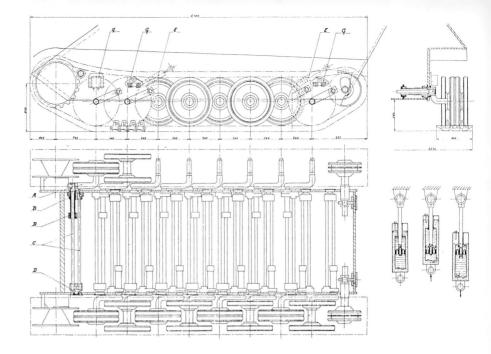

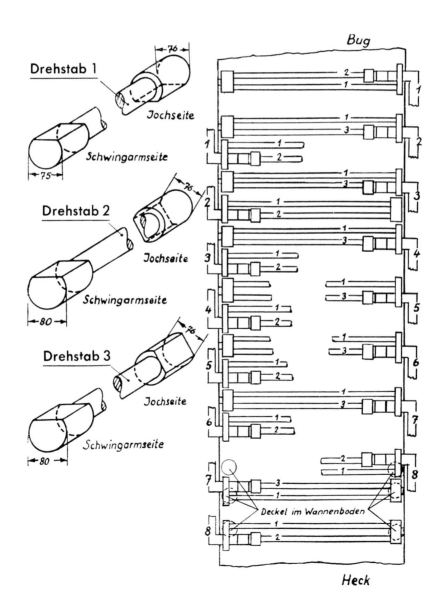

Design of the torsion bar arrangement in the Panther tank.

The quality of suspension was that much better when the amplitude of vertical displacement was smaller and less acceleration was built up when crossing broken terrain. During comparison tests all vehicles demonstrated nearly the same pitch resonance at a speed of 5 km/h. With an increase in speed the pitch resonance also increased correspondingly — with the exception of the Panther V1 test vehicle equipped with shock absorbers on the first and last road wheel arms; a steady decrease was shown with this suspension. As a consequence, this avenue would have to be explored further in order to reach the targeted goal. As opposed to the Panther V1, it was not possible to mount the shock absorbers on the first and eighth swing arms with the series Panthers, resulting in a deterioration of the flotation qualities. This was lower because the shock absorbers were mounted on the second and seventh road wheel arms; an improved flotation such as found on the Panther V1 demanded that the shocks be affixed to the first and eighth arms. This positioning of the shock absorbers apparently played a major role in the flotation characteristics, since the suspension on both Panthers was exactly the same.

Hemscheidt developed a prototype shock absorber for the Panther which demonstrated a greater volume. Compared to the HT 90 mass-produced shock absorber, the following differences were noted:

	HT 90	125
length fully extended	605 mm	710 mm
length fully compressed	445 mm	550 mm
travel movement	160 mm	160 mm

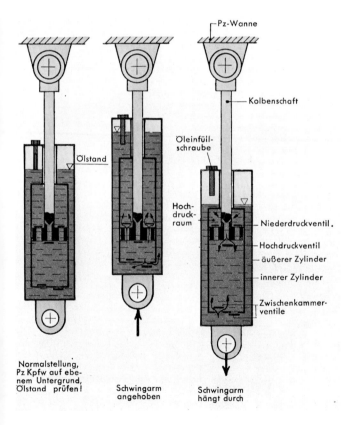

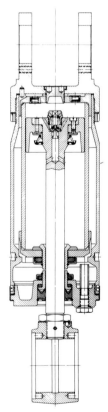

Left: Operating characteristics of the shock absorbers.

Details of the Panther's HT 90 shock absorber.

The other critical element was shaped by the two versions of the Panzerkampfwagen IV, one of which utilized a volute spring suspension. This vehicle exhibited a dramatic rise in vertical movement with an increase in speed. At a speed of 22 to 25 km/h the vibration became so great that it was not possible to maintain any higher rate of speed.

The characteristics of the leaf-sprung VK 3001 (DB) and the US Sherman M4 were approximately the same. The acceleration and impact were several times greater than with the Panther. Even with the T-34, the pitch became so great at 25 km/h that a further increase in speed was not possible. In all the Panthers tested the natural frequency of the rebound travel was significantly less than the natural frequency of the bump travel rate. At 31/min the Panther's was by far the lowest. Because of space constraints it wasn't deemed possible to bring the rebound travel's natural frequency much below 30/min.

The importance of damping cannot be overemphasized. If it was too little, a very uncomfortable "porpoising" would result at the critical speed. If it was too great, an unpleasant feeling of acceleration would result in the vehicle. It was necessary to keep the natural frequency of the vertical displacement as low as possible in order to maintain a good suspension. At the same time it was essential to pay attention to the arrangement of the shocks so that the vertical motion would be dampened at $D = 0.4$ to 0.5. Any less damping would be insufficient. If the flotation didn't meet the proscribed guidelines, then even a soft suspension would be useless. The damping was just as important as the suspension and both were inseparable from each other. During the development of the Panther the Panzerkommission expressly specified that the springs would be required to have a fatigue life of 10,000 km. The lower dissipation limit of torsional fatigue was $+- 20$ kg/mm^2 for a shaft having a diameter of 50 to 60 mm. This ensured the best application of material, especially once the fatigue life was increased to nearly double that of a simple ground shaft surface by using either surface pressurization finishing or bombarding the surface with steel bearings (shot peening/cloudburst hardening). A recurring disadvantage was that, in view of the given resistance data, a road wheel stroke of just 200 mm would require a shaft length greater than the width of the tank hull. Many options were explored with the purpose of alleviating this drawback; these involved connecting several torsion bars in tandem or parallel. Of all these possibilities, however, only the "hairpin" design as planned for the Panther led to a feasible solution. Initially

Mounting of the shock absorbers on the interior hull wall.

this design was quite a risk, since the spring bar was not only subject to torsional stress, but was also bent to a certain degree. After extensive testing showed that this double load would be carried remarkably well, this suspension could be introduced on production models.

There were virtually no complaints during Panther operations. The two torsion bars were laid one behind the other and linked by a coupler. The torsion bars had a circular nut with a milled support surface, since the mountings of Dr.-Ing. Lehr's test vehicles having this feature showed that it also prevented the bars from being incorrectly installed into their receptacles. The coupler had two cylindrical openings to accept the torsion bar nuts. The two torsion bars were held in place by shims having opposing inclined bevels. A trunnion was affixed opposite the torsion bars to the coupler and secured it into a bore of a support on the hull wall, giving the unit a pendulating motion.

A full margin of reliability could only be attained once the coupler had been strengthened. Since the installation of the double torsion bars was linked to a certain uneasiness with regards to maintenance, there was naturally the desire for a more simplified solution. Such a solution, however, could not be found. There was no purpose in implementing recommendations which only resulted in a significantly less stroke. The difficulty was namely in providing a long-lasting spring which permitted a road wheel stroke of more than 500 mm. The effective stroke was reduced to 220 mm when using a single torsion bar. In addition, this resulted in a substantial increase in the vehicle's natural frequency. Both of these characteristics meant a step backward in the current state of Panther development — a measure which was out of the question. Nevertheless, for reasons of material shortages a single torsion bar suspension was planned for the follow-on Panther II.

Due to less constructive use of materials all other types of suspension — coil spring, leaf, volute spring, and rubber — would have meant a significant increase in the Panther's weight in order to provide the same quality of suspension. The torsion bar suspension arguably provided the most effective utilization of materials. In spite of this fact, the torsion bar suspension was repeatedly subject to harsh criticism. Some areas of complaint were:

— the high amount of boring needed to be done on the tank hull and the associated high production costs. However, the suspension was armor protected and was largely protected against dirt and damage.

— the arrangement of the torsion bar suspension required a 50 to 55 mm greater hull height and total height. The front drive meant a further increase in the total height by roughly 140 mm to 195 mm due to the drive shaft channel.

— the manufacturing and refining of the swing arms formed a bottleneck. Once the arms were assembled from several parts

— some of which were tubular — using electric buttseam welding, this problem was also remedied.

— the springs crisscrossed the floor of the hull like a grill, making a floor escape hatch impossible.

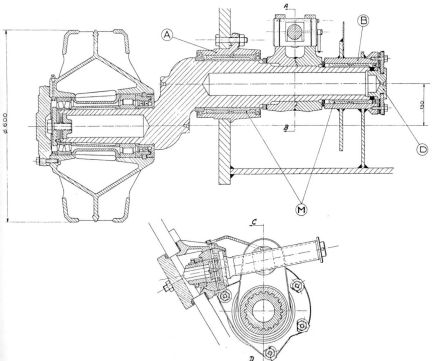

Idler wheel with spindle for adjusting track.

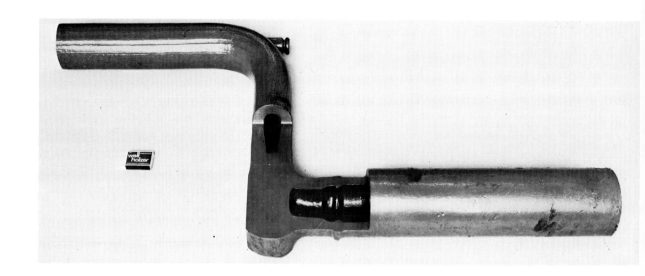

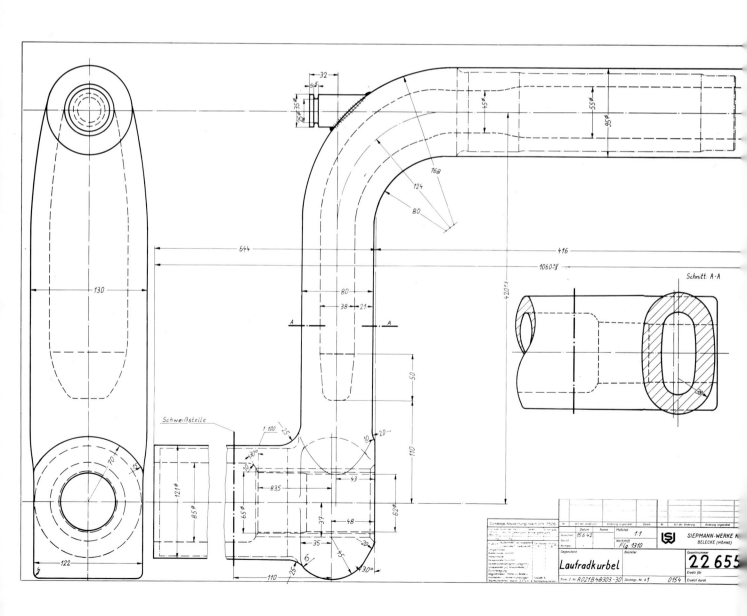

The hollow forged road wheel arm of the Panther, in photo and drawing.

Design of the hollow forged swing arm: the swing arm for the road wheels had two shafts. One (1) supported the road wheel, while the other (2) was anchored in the tank's hull. In order to keep the weight of the swing arm down while at the same time providing the best durability, the piece was made out of high strength rolled steel — hollow forged where possible. The road wheel axle (1) and the arm (3) were made from a solid stretched cylinder of steel, hollowed out using forging machines. The hub (4), which accepted the torsion bars, was affixed to the lifting arm and hollowed out on other forging machinery. The tube ending, which formed the mounting pin for anchoring the arm inside the tank hull, was electrically welded to the hub (4) using butt welding machines. The final shape of the swing arm was achieved by bending the road wheel axle (1) 90 degrees.

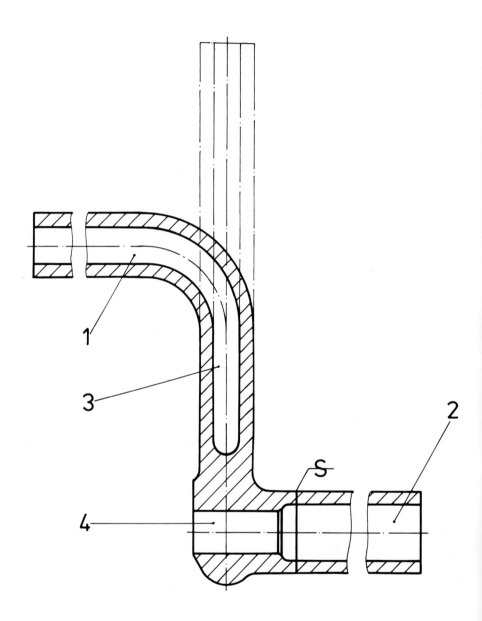

— replacing the double torsion bars was a tiresome and difficult task, particularly after being damaged by a mine.

One thing was certain, however — at this point in time there was no other type of suspension which could match the same qualities of the double torsion bar suspension.

Each side of the running gear of the Panther consisted of four inner and four outer road wheels, eight swing arms, eight double torsion bars, two shock absorbers, an idler wheel with an adjusting shaft for track tension, a return roller, three bump stop for the swing arms and one track.

Dr.-Ing Lehr had hoped to avoid stop buffers on the two forward wheels, but it later became necessary to mount these. The installation of such buffers was difficult due to the shortage of space with an interleaved running gear. The effect of these cup spring buffers was seriously reduced when they were gummed up by mud and clay. Since on one side of the running gear the road wheel arms were installed offset from the other side, a buffer had to be affixed at the first axle due to the disruption of the suspension.

The role model for the Panther's running gear was most certainly the interleaved running gear of Germany's half tracks, which enjoyed the following advantages: the large road wheels move relatively little even while traversing rugged terrain and maintained drive power when overcoming ground obstacles. The wear on the rubber tires was also kept within acceptable limits, thanks to a relatively wide grip area on the wheels and little compression of the rubber. The small interval between the eight road wheels per track gave them a relatively equal ground pressure which would otherwise not have been possible.

The bogie wheels were arranged in an overlapping manner and were dished disk wheels having removable all-rubber tires. A wheel rim was fitted to the wheel disks (interval between disks was 560 mm) for accepting the rubber tires, the second rim being bolted on. The rubber tire for the wheel was laid between the two rims and held in place by tightening the bolts (16 on the Ausf. D, 24 on Ausf. A and later models). This obviated the need for the expensive process of cutting grooves prior to vulcanizing the rubber to the wheel. Steel wire was vulcanized into the inside of the tire to retain some of the cohesion which had

Road wheels and swing arms seen from the inner side.

Details of the road wheels on an Ausf. D.

The solid rubber tires were not vulcanized to the road wheels and could be changed without using any special tools. A sledgehammer was sufficient to knock the rubber tire and retaining ring from off the wheel.

The tensioner for the tracks was located in the rear of the vehicle.

Three firms were involved in manufacturing the torsion bar springs: Röchling, Dittmann-Neuhaus and Hoesch. Whereas Röchling and Hoesch used shot peening (a Röchling patent), Dittmann-Neuhaus instead made use of a OBERFLACHENDRUCKEN (surface pressuring) process. Raw materials for the springs had a quality standard of 50 Cr V 4, but there had to be a balance reached in which the most critical alloy components such as C and Cr were kept at their maximum limit.

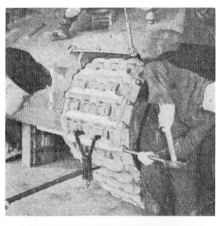

A special tool called a track connecter enabled the track to be separated and fastened together easily.

been exerted on the rubber vulcanized to the previous steel tires. This method proved to be quite effective. The inner road wheels carried track guide rings. Having a diameter of 860 mm, each wheel rested on its swing arm on tapered roller bearings. The road wheel arms in turn pivoted inside molded Bakelite bushings inside the tank's hull. The ground pressure of the bogie wheels with tires amounted to 141 kg/cm. The double torsion bars used for suspension were secured in place by shims. The ends of the original torsion bars were slotted, or keyed, according to the vehicle standards of DIN Kr 231 Bl. 1, having a radius of 0.4 mm at the base of the slot. The sharp V-angled grooves caused breaks in the torsion bars which invariably stemmed from the slots. Because of this Dr.-Ing. Lehr proposed his own design of flattening the surface of the torsion bar head and using shims to hold it in place; after extensive studies this method was introduced on production models. The flattened areas required protection from abrasive corrosion, which the torsion bar manufacturers achieved by copper plating them.

Nevertheless, the troops had serious difficulties — even if most of the breaks now occurred in the shaft and no longer at the head. The Porsche company therefore developed its own serration design with a slot base of 0.8 mm. Disregarding the advantage of having a lower notch effect, it was now possible to use the process of shot peening on the shaft as well as the slot bases, thereby achieving a higher fatigue life.

Using a cable the track was drawn over and lifted onto the drive sprocket.

The loose track was then joined together.

Track details of the Panther tank.

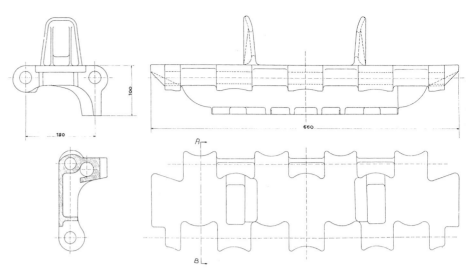

Panther seen traversing the "Wellenbahn", or undulated test track at the testing grounds of Maschinenfabrik Augsburg-Nürnberg. On the vehicle's right side can be seen the recording device for measuring the bounce (vertical movement of vehicle).

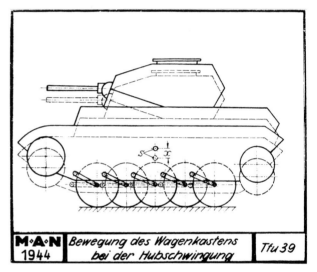

Schematic drawing showing the vehicle's body movement during bounce.

Details of the Panther's recording apparatus for measuring bounce.

Resonance curve compilation for the vertical center-of-mass acceleration.

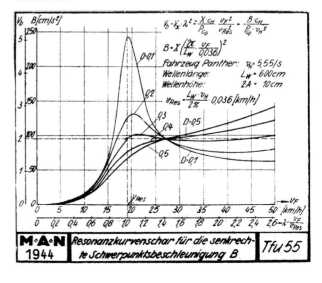

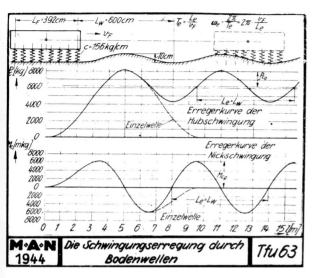

Amplitude of oscillation caused by ground undulations

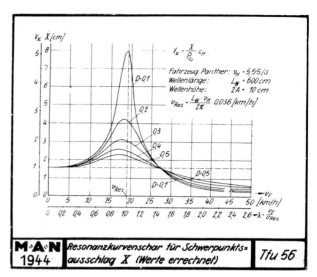

Resonance curve compilation for center-of-mass vertical displacement (data calculated)

Results of vehicle longitudinal inclination (pitching motion) of tanks at the Betonwellenband (undulated track) 2 of the Verskraft Kummersdorf. The diagram shows the excellent performance of the Panther running gear. (The track length was 1 km)

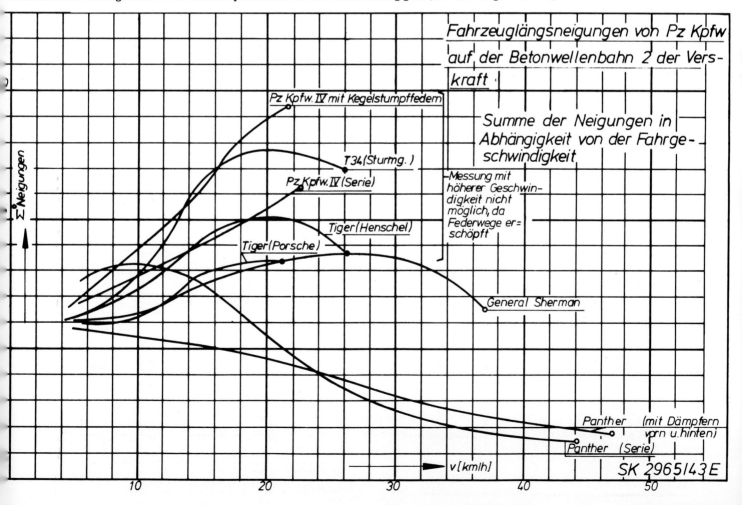

The shock absorbers could only be fitted inside the hull, since the interleaved suspension offered little room for the normal installation of the telescoping shocks. The shock absorbers were mounted above the second and seventh road wheel arms to a lever 250 mm long fitted to the swing arm trunnion on the inside of the hull wall. Original plans called for installing the shocks on the first and eighth swing arms, but due to space constraints this was not possible on mass produced vehicles. The shock absorbers functioned unilaterally and only cushioned the downward motion of the road wheel.

The difference in shock absorption effectiveness of the two arrangements was revealed on a special undulated test track of the Verskraft in Kummersdorf. The following amplitude ratios were measured over a distance of 1000 m (data in parentheses are for series vehicles having shocks on the second and seventh swing arm): 10 km/h = 220 (330), 40 km/h = 90 (65). The drop in the vertical amplitude ratios for production vehicles at increased speeds was attributed to the arrangement of the shock absorbers. The drawbacks of installing the shocks inside the vehicle were an increase in the temperature for the driver and radio operator and a major increase in the stress on the hollow trunnions of the swing arm. This meant that the second swing arm could not be manufactured hollow like the others, since the hollow arm could not accept the bending and torsional forces from the wheels without becoming warped. The 420 mm long hollow swing arm was a special feature of the running gear. The angled swing arm for the bogie wheel and the crosspiece hub in which the torsion bar fit were both made of forged metal onto which the hollow trunnion running into the tank interior was welded. Teething troubles resulted when greater demand later brought the Eisenwerke Oberdonau in Linz into the production process; the wall of the inner hollow swing arm was of varying strength due to the hollowing process. The hollow swing arms were a development of the Grossschmiede Siepmann, Beleke in Sauerland.

The rear-mounted idler wheel was fitted on a cranked axle resting in bushings inside the Panther's hull. Tensioning the track was accomplished by drawing the idler wheel axle back using left turns on the adjusting spindle of the track tensioner.

Via lubricating lines, two centrally-located batteries provided lubrication to the swing arm bearings, the idler wheel bearings and the shock absorber pins. The single-link single-pin tracks (Type Kgs 64/660/150) consisted of 86 links joined together by non-lubricated pins. The track pins were held in place with spacers and expansion pins. The specific ground pressure of the track was 0.87 bar. The developer and manufacturer of the tracks was the firm of Ritscher, Moorburger Treckerwerke in Hamburg-Moorburg.

The disadvantages of this type of running gear became readily apparent during the muddy periods in Russia; during sudden frost the heavily clogged overlapping road wheels would become frozen together if the mud had not been cleaned off beforehand. Shell damage would cause the bogie wheels to jam together; separating them could only be accomplished with great difficulty. In addition, a single damaged wheel often necessitated the removal of several. For these reasons this technologically advanced running gear design was never built again following World War II.

The electrical system had a current of 12 volts. Two 12 volt lead batteries, each having a capacity of 120 or 150 Ah, were installed in the fighting compartment beneath the floor plates. The generator with a separate voltage

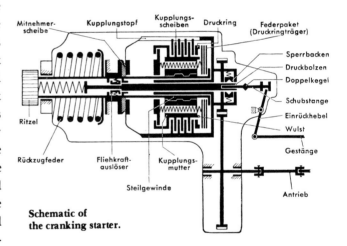

Schematic of the cranking starter.

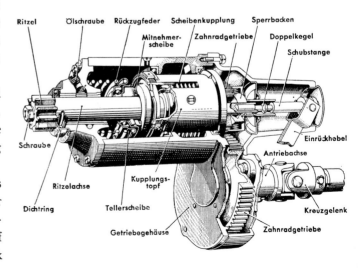

regulator provided an output of 700 watts. Cooling air for the generator was drawn in from the fighting compartment through a metal tube. The electrical system was fully grounded.

In addition to the electrical starter of 24 volts, two other means were available for manually starting the Panther's engine: the crank starter, Type AL/RBI/RI or the inertia starter, Type Al/ZMI (both made by Bosch). These made it possible to start the vehicle independently of the internal electrical system of the tank, particularly useful during extremely low temperatures (a requirement at -40 degrees C). The inertia starter was a compact design which, like the electric starter, was fastened to the engine and engaged the engine flywheel with a pinion. The starter flywheel was set into motion by a hand crank via a planetary drive and spur gear reduction having a ratio of 140 to 10,000 1/min. Transferring this stored kinetic energy was accomplished by mechanically meshing the pinion with the teeth on the engine flywheel and then coupled to the starter flywheel by means of disk plates. In order to ensure a secure mesh this plate coupling was done in steps, so that although the pinion was already spinning slowly the connection was only locked after meshing. If a certain r.p.m. was exceeded the disks would slip on each other, thereby providing a way of absorbing the force. Additionally, the coupling was also free running, meaning that when the pinion was overtaken by the engine's flywheel r.p.m. during startup the crank starter was prevented from being radically accelerated by the engine. After the engine has been started a return spring pulled the pinion out of mesh.

For the first time ever, an automatic fire extinguishing system was used in the Panther. If the temperature exceeded 160 degrees C a special extinguishing mixture of carbon-chlorine-bromine (CB) was sprayed on the hazardous area for approximately seven seconds. The system initially functioned quite poorly. The system's heat sensors were designed to be cooled by the stream of mixture. But since the spray nozzles and the sensors were not held in place by the same mount the sensors were inadequately cooled and set off another burst from the extinguisher. A joint mount improved the degree of effectiveness of the system. Another weakness, however, could not be rectified. Three heat sensors were insufficient for monitoring the temperature in the engine compartment. Even if there were no fire, the system would automatically be activated by the high temperatures in the engine compartment. This meant that only a reduced supply was available in case of an actual fire.

The situation was improved somewhat by only having two extinguisher bottles activated automatically; the remaining two could be operated by hand.

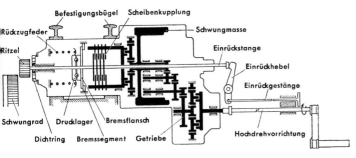

Schematic of the inertia starter (Bosch).

Cutaway of the inertia starter.

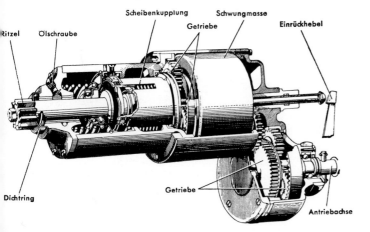

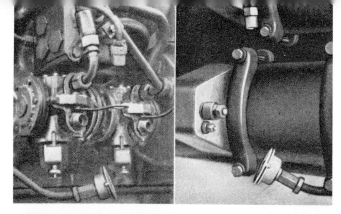

Overview of the automatic fire extinguishing system.

The automatic fire extinguishing arrangement made use of a heat sensor and an extinguisher at the carburetor, at the fuel pumps and at the starter.

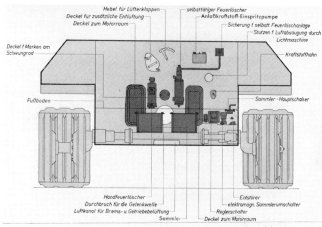

The extinguisher containers were mounted on the firewall between the fighting compartment and the engine compartment.

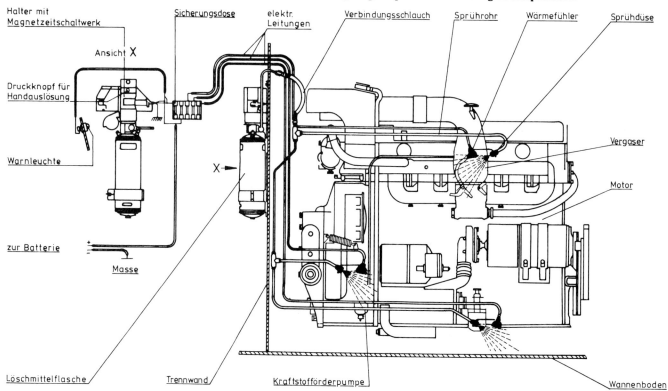

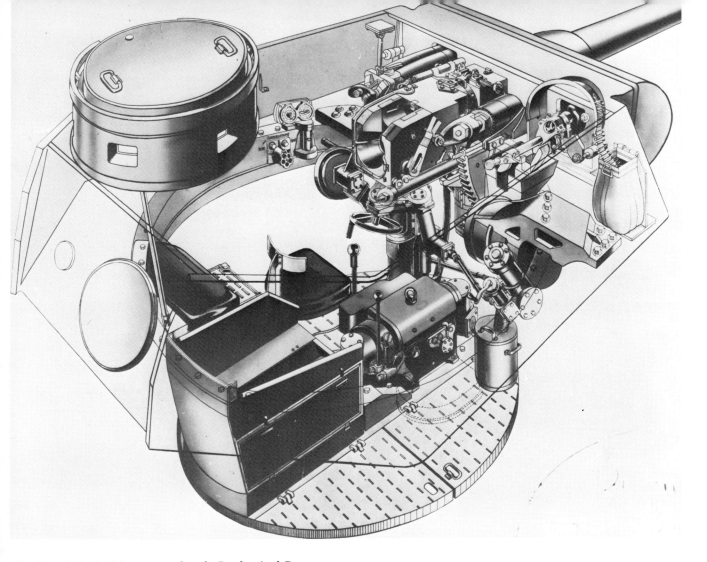

See-through sketch of the turret used on the Panther Ausf. D.

The three-man turret rotated through a full 360 degrees and supported itself on a carrier ring resting on the turret bearing race of 1650 mm diameter. It was composed of the turret sidewalls, the turret roof with commander's cupola and the mantelet with main gun and machine gun. On 18 July 1941 a contract was awarded for the development of a cannon with a penetrating force of 140 mm at a distance of 1000 meters. This resulted in the 75mm KwK 42 L/70, built by the firm of Rheinmetall-Borsig AG in Unterlüss. The vehicle carried 79 rounds of ammunition for the main gun. The 75mm Kampfwagenkanone 42 L/70 was a semiautomatic weapon with fired electrically. It fired solid core ammunition and high explosive shells (cartridges). The weights and dimensions of the rounds were: Panzergranate (AP) 14.3 kg, 893 mm long; Sprenggranate (HE) 11.2 kg, 929 mm long. The targets were sighted through the Turmzielfernrohr (TZF) 12 having a 2.5x enlargement and a 28 degree field of view. The barrel was of continuous design with a removable breech plate. It was mounted on and adjusted by the cradle. A baffle was affixed to the muzzle end of the barrel and acted as a muzzle brake countering the recoil of the barrel. During firing, roughly 70% of the recoil was absorbed when the shell's explosive gasses struck against the baffle's break plates inside the muzzle brake, were directed out the sides and thereby acted against the recoil motion of the barrel. The gun could not be fired without its muzzle brake. The breechblock was mounted on the breech plate as were the gun sighting mechanisms. The gun barrel was equipped with a type of dust extractor, operated by compressed air. The cradle supported the barrel, which also gave it the direction of fire by the elevation and traverse controls. The recoil brake was driven hydraulically while the counter-recoil was pneumatic. The turret basket was securely fixed to the turret. A slip ring transfer was used to relay the electrical power from the chassis to the turret. The turret was sealed against the tank hull with labyrinth packing. The small gap between the carrier ring and the outer bearing race could be sealed off against water penetration by a rubber coated metal ring.

During road marches the gun was held in place with a barrel lock, which folded down onto the forward hull

Turret, left side.

View into the turret, from front.

Turret, right side.

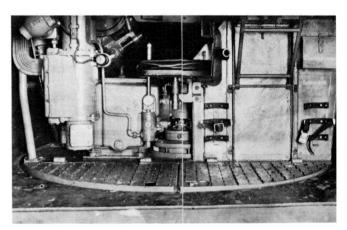

Turret basket, left side.

when not in use. The gun was elevated by hand using the elevating mechanism. The turret traversing mechanism served to adjust the gun's traverse and was operated manually or hydraulically. Hydraulically operated, the turret could make one complete 360 degree revolution in a minute.

The pressure needed to traverse the turret was produced by a hydraulic drive, which was driven by a cardan shaft from the turret drive. During road marches the hydraulic drive could be disconnected by a lever from the loader's position by means of a jaw clutch coupling mounted in the same housing as the drive. Manual operation was accomplished using the hand wheel of the traverse mechanism, the rotation of which was also transferred to the shifting clutch via a transmission unit. An auxiliary drive on the right side of the turret enabled the loader to assist the gunner when manually rotating the turret. The hydraulic drive, a Boehringer-Sturm Type M 4 S, operated independently of the engine r.p.m. and consisted of two vane-type units with rotating housings; they were both of the same design, but one was driven as the pump and the other functioned as the hydraulic motor. They were joined together in an enclosed cycle by a suction and pressure channel inside a fixed tubular body. Regulating the drive r.p.m. of the hydraulic motor, i.e. regulating the traverse speed of the turret, was controlled by monitoring the discharge from the pump.

For sighting the gunner used the TZF 12 binocular sight through two vision ports in the gun mantelet. The gun elevated from -8 degrees to +20 degrees. On the first prototypes the commander's cupola was cut into the left side wall of the turret. On series production models it was shifted more to the center of the turret. On older Panther Ausf. D models a circular communication opening was cut into the left turret wall. Additionally, these turrets also had a pistol port in each side and rear wall which could be sealed with a plug. The turret traverse indicator ring dial in the Ausf. D had two numbered gauges; the left one was divided from 0 to 12 on the inner scale and 0 to 64 on the outer scale. It served for coarse adjustments. The right gauge was had a linear division from 0 to 100 and was used for fine adjustments. The turret basket on the Panther Ausf. D consisted of angle iron and plating. It functioned as the floor plate for the gunner and loader (each having a space of 570 mm). The commander's cupola was comprised of a cylindrical housing, the aperture block, the

Gun with armor mantelet and base plate.

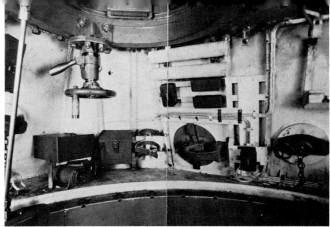
Interior view of turret left side wall.

Turret basket, right side.

Turret bearing race with seal.

hatch and the clock-dial turret traverse indicator with control adjustor. On some vehicles, multiple smoke grenade dischargers were fixed to both sides of the Panther Ausf. D turrets. Later, a close combat weapon, or Nahverteidigungswaffe, was installed in the roof of the turret. This weapon could fire the Nebelkerze 39 smoke grenade, mortar rounds, smoke markers orange 350 and Leuchtgeschoss R star shells.

For self defense the Nahverteidigungswaffe could fire 2.6 cm HE shells through the use of a flare pistol. These had a range of 7 to 10 meters and would detonate 0.5 to 2 meters above the ground; the fragmentation effect reached to 100 meters.

The initial Panthers were equipped with a simplistic, ball-shaped muzzle brake for the main gun.

5100 rounds of 7.9 mm ammunition were carried in the tank for the two MG 34 machine guns. In the Ausf. D the radio operator fired through a hinged flap instead of an integral ball mount. The vehicle carried a crew of five.

Whereas the Pzkpfw II, III, IV and Tiger tanks carried the crew's baggage in a compartment on the rear of the turrets, the stowage boxes on the Panther were mounted on the right and left of the outer rear hull wall.

Turret traverse drive with oil pressure pumps.

Exterior view of turret with plugs for pistol port and communications hatch. Immediately below the turret can be seen a hole where a shell had penetrated the armor.

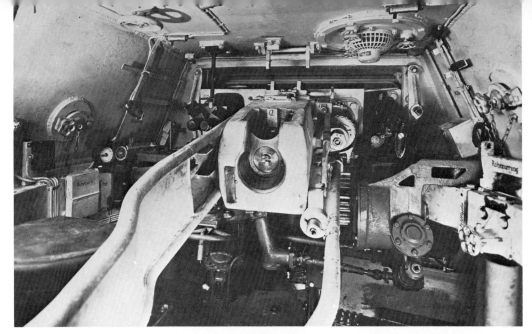

Details of the Panther turret interior on an early version. The commander's seat is to the far left.

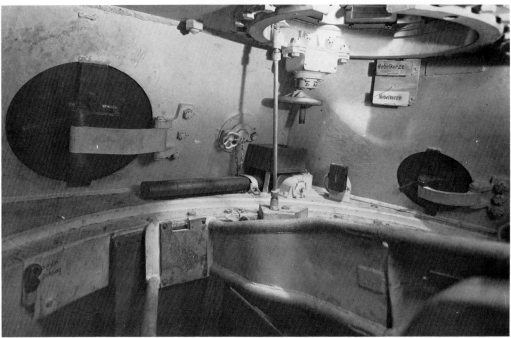

The interior of the Panther turret shows the openings for both the pistol ports and communications. The twin gauge for the turret position is just below the pistol port.

On the Ausf. D there were still twin gauge dials for the turret traverse drive.

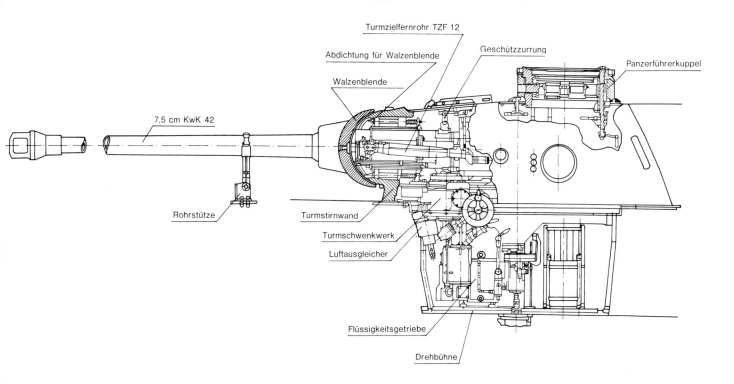

Comparative drawings of the turrets for the D series (above) as well as the A and G series. Cutaway drawings are of the left turret side (sketches drawn by F. Gruber).

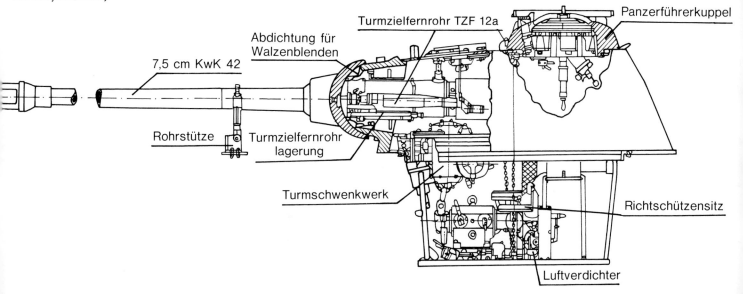

At this point it is necessary to devote a few paragraphs regarding the development of armor up to this time.

The First World War saw the birth of armored vehicle development; the so-called "skeleton" design used in these early vehicles continued in nearly all armor-building countries until well into World War II. This construction method involved riveting or bolting commonly found thin armor plating (6 to 20 mm) to a framework of riveted and/or bolted flat and angle irons which corresponded to the outer form of the vehicle. Such a design meant that the plate edges had to be accurately machined with regards to shape and mitering in order to provide seamless blending of the plates. Because of the inadequate strength of the tool metals at that time the tensile strength/temper of the armor plating had its limits. The internationally used high alloy metals having a mixture of chromium, nickel, tungsten and molybdenum could therefore only be tempered to a tensile strength of roughly 100 to 120kg/mm^2.

By 1918 the Deutsches Reich had officially produced only 20 "A 7 V" combat tanks (as opposed to more than 6000 Allied tanks), which had resulted in no new developmental thinking. If the Versailles Treaty prohibited Germany from any type of armored vehicle, it

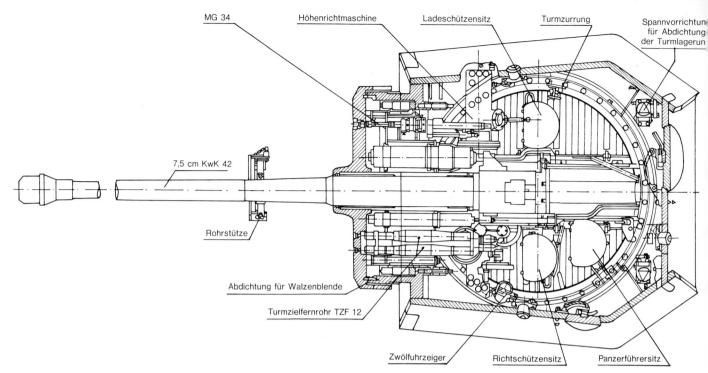

Cutaway of the turrets for the Ausf. D (above) and both A and G (F. Gruber).

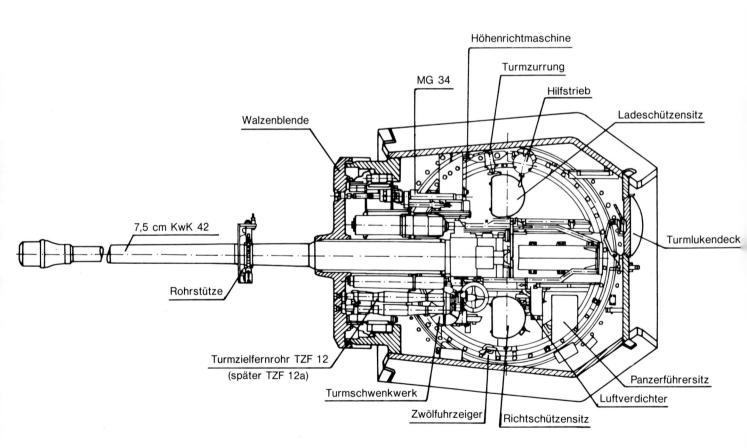

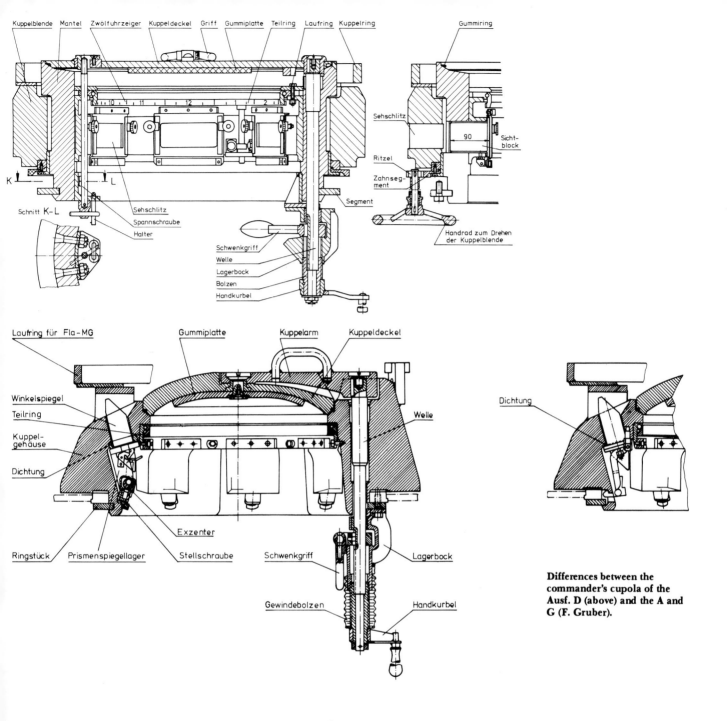

Differences between the commander's cupola of the Ausf. D (above) and the A and G (F. Gruber).

could not stop the Reichswehr and its ministry from making thorough studies of the world's tanks being developed after 1918. From these studies there evolved the technical improvements which were to find their expression in the first German armored vehicles — the Panzerkampfwagen LaS and LaS 100 (later the Pzkpfw I and II) and the Rad-Panzerspähwagen (wheeled reconnaissance vehicle) Kfz 13 and 14.

Since Germany's experience with armor up until around 1932 involved only naval applications dating back to 1918, then with metal helmets and protective shielding (meaning primarily large surface thick plates of nearly flat contours, or thin plating), it became necessary to organize an armor development program for military applications in light of the tank experience gained. This began initially with plate thicknesses of 5 to 13 mm of steel having a reduced chromium and nickel content and also with little tungsten. This resulted in a tensile strength of 180 kg/mm² once carbide tipped tools (Widia) were introduced in the manufacturing process. However, the big success here was in finding a way to make a "shellproof" seam using edge welding, thus making the entire "skeleton" construction superfluous and bringing about an armor weight reduction of 5 to 10 per cent — a very significant savings indeed. It also dispensed with thousands of rivets and bolts, which in combat were torn away in great numbers with every hit and posed a danger both to the crew and to the sturdiness of the vehicle.

It must be emphasized here that from the very beginning Germany only built tank bodies using the welded type of construction, just as from the beginning only electric steel was permitted in the melting process. By 1945 numerous improvements had been made which had been worked out and introduced on the initiative of the Panzerabteilung in the Heereswaffenamt of the OKH, WaPrüf 6 II. Under the direction of Oberregierungsbaurat Dipl.-Ing. Walter Rau, the Panzerabteilung initially worked in conjunction with first one, then four, then seven and finally 18 steel mills and 18 steel foundries in addition to several specialized firms.

By 1945 the precarious supply situation of critical alloy materials facing Germany resulted in a program of intentional systematic reduction of nickel, tungsten and molybdenum in the composition of armor steel. Given a thickness range of 5 to 13 mm the resistance was reduced from 180 kg/mm2 to approximately 150 kg/mm², since with regards to toughness (i.e. protection against breaking) it was more beneficial to have a resistance strength between 120 and 150 kg/mm². Throughout the entire thickness range — the upper limits of which remained conformal with the ever-increasing anti-armor calibers and shell types to a maximum of 250 mm — first nickel-free and then low alloy steels were introduced in compliance with carefully determined heat treating methods in the smelting of this "standard" steel for all plants. Depending on the gauge, the resistance varied

A period document, which — better than most others —illustrates the prevalent conditions of the time. It reads: Announcement:
Reichsminister Speer, of the Reich Ministry for Armament and Munition, has permitted a cost-free special delivery of pickled sardines to our establishment. Approximately 1 kg may be distributed to each German employee (including youth) from the shipment which has just arrived. In order to conduct the distribution in the most expedient manner possible, it is necessary that the employees bring their own wrapping paper from home. Time and location of distribution will be announced in the individual departments.
This donation should give us a renewed incentive to devote ourselves to fulfilling our assigned responsibilities.

Bekanntmachung

Der Herr Reichsminister S p e e r , Reichsministerium für Bewaffnung und Munition hat unserer Betriebsgemeinschaft eine kostenlose Sonderzuteilung von

S a l z s a r d e l l e n

zukommen lassen. Aus der nunmehr eingetroffenen Sendung kann jedem deutschen Gefolgschaftsmitglied (auch Jugendlichen) etwa 1 Kg zugeteilt werden. Um die Verteilung bestmöglichst durchführen zu können, ist es erforderlich, dass die Gefolgschaftsmitglieder Einschlagpapier von zu Hause mitbringen. Ausgabezeit und Ort wird in den einzelnen Abteilungen bekanntgegeben.

Die Spende soll uns erneut ein Ansporn sein, uns für die Erfüllung der uns gestellten Aufgaben einzusetzen.

Friedrichshafen a.B., den 2. Dezember 1942

Der Betriebsführer: Maybach   Raebel   Rommel

between 150 to approximately 100 kg/mm². With only a minimal drop off, the quality of steel remained at this level up until the end of the war with regards to physical composition and protection against shell damage; the deliveries of tank bodies to the assembly plants also continued at a nearly uninterrupted pace. Temporary changes in the alloy content, the rapid transfer of experience data and the elimination of errors by one or the other manufacturer, as well as bridging problems caused by war damage — all these were only possible because the producers of armor steel as well as those manufacturing cast armor were part of a "special" committee under the direction of Dir. Dr. R. Scherer (DEW Krefeld) or Dir. Dr. K. Roesch (BSI Remscheid) which worked in close connection with the WaPrüf 6 II. This office had the final say in authorizing and clearing material for the army's armor programs.

A significant reduction in work time, costs and waste was attained by 1944/1945 as the machine cutting of armor pieces was virtually completely replaced by a mechanical process of autogenous flame cutting. This initially began using oxyacetylene cutters but then switched to natural gas-oxygen for the cutting of all outer and inner contours — thereby both reducing the cost even further and providing a greater measure of operational safety. Even cuts on extremely thick armor pieces were made without difficulties using this method.

A prerequisite for this approach was ensuring that the edges which butted up against the "armor weld" seams were overlapping and notched; this technique deviated from the normal type of weld joint used in civilian steel construction. The strength of the weld and protection against shells was achieved by a two-way support of the armor pieces, so that only roughly one-quarter of the area in contact required welding. The stress load on the welds of the armor pieces (the thickness of which was constantly increasing) was alleviated through the use of tooth and mortise joints. Further progress was made insofar that the original "shellproof" weld joint, comprised of three different weld layers, was replaced by the previously mentioned "armor weld seam" which only required a single integral welding electrode. This austenitic type of electrode delivered an exceptionally tough and solid weld which held together even under a direct hit and made possible the joining of armor plating with cast armor pieces.

For several years the manufacture of face hardened armor parts remained important, until their additional protective effect over that of homogenous armor parts was overtaken by increasingly more powerful ammunition. The method used for naval armor until well into the 1920s was to coat the exterior of armor plates (for roughly 10% of the total plate thickness) with a high carbon content exceeding 1% which, depending on tempering, delivered a high degree of hardness of approx. 58 to 62 Rc in the main cross-section of the plate, giving 100 to 110 kg/mm². This was invented and called powder cementation by Harvey (USA) and further improved by Krupp. But the required recarburizing time of approximately 24 hours per 1 mm harder layer was only practical for naval purposes and was not practical for mass production of armored vehicles.

Even the somewhat less time-consuming method of recarburization with natural gas was not adequate. Truly perceptible progress was only reached in 1935 when, under the direction of the Heereswaffenamt/WaPrüf 6 II, the firm of Peddinghaus-Gevelsberg i.W. succeeded in developing linear torches for natural gas-oxygen to a width of 1500 mm. Coupled with a follow-on rinse, the corresponding face hardness was achieved as with the cementation process — only now in the same amount of output per minute as previously had been produced in hours. This flame hardening, also called the O.h. process, could provide armor thicknesses of up to 100 mm with a hardening depth of 10 to 20 mm. However, for manufacturing plants which did not produce their own steel, i.e. didn't have their own supply of natural gas, the electro-inductive surface hardening process was developed at around the same time. This used either medium or low frequency current, depending on the thickness of the material, and to a certain extent was even more efficient than flame hardening. Both processes were effective on even very heavy pieces. The technological advantage in shell protection/armor thickness was remarkable not only primarily against standard armor shot, but also against small caliber solid core projectiles, and remained so until the advent of the hollow charge round. Finally, the progress attained in armor body structure is worth mentioning not only because of the shell protection it afforded, but also because of its importance to a more economical production output.

Continuing on from the self-supporting welded construction design, a lower piece (or hull) was joined to the body along with the one or two superstructure armor pieces bolted on. On this rested the traversable turret or, on Jagdpanzer and self-propelled vehicles, the fixed superstructure. The elimination of the undesirable bolting of the main components was achieved by use of manufacturing equipment for rotating the vehicle and for

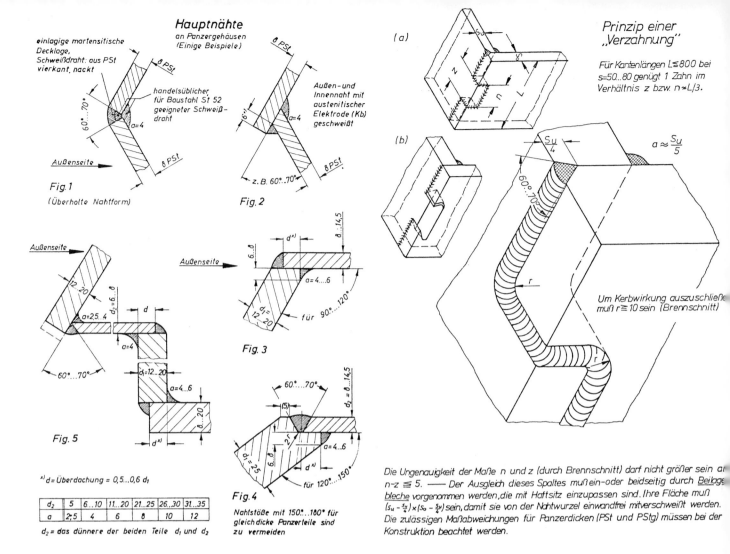

Illustration of weld seams for armor plating.

Principle of toothed interlocking for armor plating.

component welding. In doing so, the nearly vertical glacis as found on the Panzerkampfwagen I to IV and even the Tiger Ausf. E was finally dispensed with (although the latter tank utilized welding instead of bolting for the front plate). It was now possible to produce a continuous sloped glacis such as on the Panther, Tiger Ausf. B, Jagdpanther and Jagdtiger. Unfortunately, the timely conversion to this type of form was dependent upon the senior-ranking leaders, who forestalled the ongoing conversion of manufacturing facilities and the establishment of the necessary equipment at a critical juncture. One could get the impression that the necessity for this improved body form was only felt by the Germans after the appearance of the Soviet T-34.

However, those who understand the time span required for changing over to such a fundamental type of construction and machining materials will also know that the interval between the T-34 and the Panther was too short for spontaneously implementing such a conversion. In reality, the necessary detail studies had been conducted long beforehand, so that the results were ready for application once the T-34 made its appearance.

As a result of previous research, the principle of spaced armor was employed by the Germans beginning in 1942. At certain critical points "supplemental armor" was placed with a space separating it from the integral armor of the vehicle and acted as a counter to hard core armor piercing and the later hollow charge projectiles. Spaced

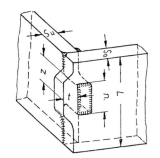

Prinzip einer „Verzapfung"

Für Kantenlängen $L \leq 1000$ bei $s \geq 80$ genügt 1 Zapfen im Verhältnis z bzw. $n \approx L/3$.

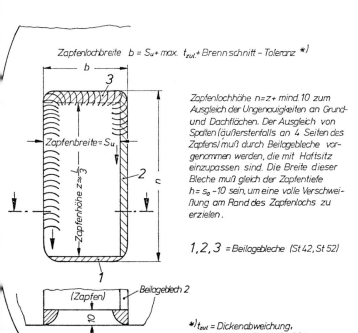

Principle of mortise interlocking for armor plating.

armor was commonly found on the Pzkpfw III and IV; the latter, along with the Sturmgeschütz assault guns, the Panther and the Tiger, also made use of protective track armor called Schürzen (skirts). With the exception of Schürzen, the Panther and Tiger tanks did not utilize any other type of spaced armor, since from their inception their armor already provided the greatest protection then available, namely the equivalent of 150, 200 and in places 250 mm. Although it was a makeshift application, the supplemental spaced armor — whether made from simple steel plating or even wire mesh, or the later armor plating — proved extremely effective against hollow charge projectiles and magnetic charges and produced an abundance of valuable knowledge.

Up to 1945 vision blocks, such as the periscopes, were considered as part of the armor and were therefore included in the overall design of vehicles. The shell-proof glass blocks were made for the Deutsche Wehrmacht exclusively by the firm of "Sigla" Sicherheitsglas GmbH in Weisswasser. An example is the "Glasblock 90", used in the Panther tank (among other vehicles), which initially had a multi-layer arrangement of crystal glass plates in thicknesses of 4+20+20+20+4 mm. Together with two soft plexiglass plates the block was 70 mm thick. A 10 mm thick inner plate was set 10 mm behind this 70 mm glass block. The resulting hollow space was designed both to trap glass fragments before reaching the crewman's eyes when the outer block was destroyed as well as to prevent the block from misting up. The "Glasblock 90", with its 70 mm front combination, was shell-proof against pointed steel-core armor piercing rounds. When enclosed in an armor steel "box" the observer was protected against angled shots. The "Glasblock 90" (with frame) could easily be removed and replaced by releasing a "beer bottle" type clasp.

On 3 February 1943 Henschel reported that only 44 of the planned 201 single-radius epicyclic steering units had been manufactured, an announcement which seriously jeopardized Panther output.

A telegram sent by Kroemer (the individual responsible for the Panther program at the Speer Ministry) to Henschel on 8 February 1943 is indicative of the hectic pace prevalent during the production start-up phase: "Why are only five February vehicles on the assembly line when ten were to have been produced in February? Why were the five January vehicles not accepted and delivered by 7 February as agreed, and when will this now take place?"

Henschel blamed the delays on problems with the delivered component parts. Additionally, it was hardly possible to assemble vehicles when the plans for the supplied components are received the month following the delivery date. Of the five Panthers in the road test department, three of them had been brought up to the latest standards and the other two were still in process. During road tests, however, the first three vehicles were revealed to have so many defects that none of them could be accepted.

A conference at MAN on 5 February 1943 resulted in a decision that sometime after the 17th MAN Panther was completed the clutch-and-brake steering would be replaced by the single-radius epicyclic steering unit. At this time the hull deck was still uneven to the point that each separate turret required individual, precise fitting, either by adjusting the bearing race carrier ring or by inserting a large spacer ring between the turret and race.

The development of a rotatable and tilting driver's periscope was to be undertaken at all costs.

The Maschinenfabrik Augsburg-Nürnberg agreed to produce five tank turrets in February of 1943 for the planned Panzerbefehlswagen (command tank) Panther, with another seven to follow in March.

In a longer meeting on February 16th, 1943, the subject was broached as to whether Italy should license-manufacture either the Panther I or the planned follow-on Panther II.

It appeared, then, that a follow-on model to the Panther had already been anticipated even though the start-up difficulties of the Panther I had in no way been eliminated. It was planned, in the spirit of mutual cooperation with Italy, to have that country send manufacturing teams to Germany while Germany, on the other hand, would send certain teams to Italy for the optics, electrical systems, etc. Another proposal was to simply deliver completed tanks to Italy in exchange for gun barrels and armor plating.

It was determined at a meeting in the Italian War Ministry in Rome on 22-24 February 1943 that Panther production would begin one year after Fiat-Ansaldo received the plans. This would allow Fiat to reach a monthly maximum output of 50 vehicles 18 months hence; as soon as a batch of more than 25 tanks had been built in a month it was to be delivered to Germany. At this point in time Fiat had much of its machinery out of operation. The Italians were made aware that beginning in the autumn of 1943 production of the Panther II was scheduled to begin. The Italian General Staff indicated that it was in agreement with the German proposals, but nothing more resulted from the plans to manufacture Panther tanks in Italy. MAN was also directed to supply Japan with a complete set of plans and an sample vehicle for the purposes of license-manufacture. However, as with the Tiger tank the war situation prevented the Panther from being delivered. Rheinmetall announced on February 18th 1943 that, according to schedule, a turret traverse drive linked to the engine r.p.m. and a simplified elevation control would be installed in all vehicles after number 851 from July 1943 on. In addition, a decision was reached to produce the new commander's cupola with periscopes from cast steel armor. It was strengthened to a minimum of 100 mm thickness, and the plans for the new cupola were to be delivered to the tank manufacturing firms by February 27th, 1943. During a demonstration for Minister Speer on 21 February, twelve Panthers with clutch-and-brake steering were supplied to the troop training grounds at Grafenwöhr. Speer criticized the hard steering, since apparently the steering brake gripped too soon. Both standing and towed targets were fired upon, but due to inadequate turret ventilation only a few rounds could be fired when the turret hatches were closed. Nevertheless, both the minister and the troops appeared enthusiastic with the overall conception of the vehicle.

In March of 1943 a contract was awarded by the Heereswaffenamt/WaPrüf 6 to J.M. Voith GmbH in Heidenheim (Brenz) for the development of a hydrodynamic transmission for the Panther and Tiger tanks. This transmission was to be installed in place of the mechanical and semi-automatic transmissions previously used, would occupy the same space and would be coupled to the existing steering units. Performance was rated at 650 metric hp, with a maximum speed of 45 km/h. The transmission consisted of two torque converters, the filling and draining of which operated the converters and acted as the shifting process. The torque converters could be run in four gears in conjunction with a mechanical two gear transmission. The two-way shifting involved the power flux of each converter being transferred to the gears in the transmission in such a manner that when one converter was filled and in operation, the gear change of the non-operating converter was carried out load-free by engaging mesh couplings. The actual shifting took place without an interruption in the drive power.

All design plans and manufacturing documents were prepared for this hydrodynamic drive. However, the practical use of the transmission was anticipated for the E-series tanks, for which a hydrodynamic transmission had been planned from the very beginning. This would make possible a more favorable, integrally suitable drive. Development was carried out under the Voith codeword "Panta."

The planned development of a hydrodynamic transmission for the E-series Type E 10 and E 25 tanks involved the transmission and steering unit, together with the cooling system, being formed into an integral, compact unit mounted in the rear of the vehicle. The Voith codeword for this program was "Arta." Parts for a prototype edition were being manufactured by the war's end.

On 23 March 1943 in Nuremberg MAN, together with representatives of the industry, determined the scope of modification work which would be undertaken jointly at the Reichsbahn facility in Berlin-Falkensee (which had the requisite heavy lift equipment and machinery). Among other things, the most significant changes would involve: installation of two new air exhaust pipes in the engine compartment, replacing the fuel lines, replacing the fuel tanks having defective weld seams, improving the lubrication of the planetary gears in the steering mechanism, installing new spur gears and pinion gears (double gear with module 12) in the final drive, changing the attaching method for the drive sprockets, changing the attaching method for the brake housing, replacing the track pins.

The second and seventh swing arm were to be replaced by a stronger type. The torsion bars on the first ten vehicles (MAN 6, DB 2, MNH 1, HS 1) were to be replaced because the wedge surface points did not correspond to the latest version. In case the necessary changes had not yet been made to the transmission, these were to be undertaken by the firm of ZF. The single-radius epicyclic steering units installed in the series models of the Panther had defects which stemmed primarily from poor oil seals. These were to be eliminated by design changes to the housing and installation of new oil gaskets.

All companies participating in the manufacture of the Panther were compelled to send foremen and mechanics to Berlin-Falkensee for the modifications in order to make these vehicles suitable for front-line action as quickly as possible. The rebuilding contract in the Reichsbahn Berlin-Falkensee repair facility was awarded by the OKH/Heereswaffenamt WuG 6 and lay in the hands of Demag Fahrzeugwerke GmbH. Daimler-Benz received a contract from the Heereswaffenamt for the technical supervision of the undertaking as well as advising the firm of Demag in all pertinent commercial matters.

On 31 March 1943 Henschel was informed that MAN and MNH were to make the changes in the steering units on the vehicles which they themselves delivered, while Henschel was responsible for the single-radius epicyclic units installed by Daimler-Benz as well as their own units.

The loss of mechanics to the planned Falkensee rebuilding program posed serious problems for all facilities involved since there was a shortage of skilled labor everywhere. The demand for the replacement parts needed for the program had reached such a high level by mid-May 1943 that it dangerously affected the production at individual plants. The organization of the rebuilding program left something to be desired; improper disassembly rendered many parts unusable. There was no division of vehicles based on manufacturer. Of the 108 Panthers in the Berlin-Falkensee Reichsbahn repair facility, only five were ready for delivery on the scheduled deadline date. Additionally, some experts felt that the quality of the overhauled vehicles suffered from carelessness.

For various reasons many of these collected Panthers suffered breakdowns, as did a large number of vehicles of Panzerabteilung 51 and 52 at the Grafenwöhr troop training grounds. Once again, the manufacturers were called upon to give up mechanics and technical personnel to provide assistance to the troops. At the end of May in 1943 Hitler was made aware that the Panther was still suffering from teething troubles. It would have been risky to deploy these vehicles in the Southeast (Greece), since problems there could not be alleviated as rapidly as in front areas closer to the Fatherland.

The single-radius epicyclic steering repeatedly proved to be a bottleneck and even temporarily brought production to a standstill at a few manufacturers. In order to overcome this obstacle as quickly as possible it was arranged that MIAG would deliver five of these units in August 1943 and 25 in September. At the beginning of June 1943, Henschel began using a Panther Ausf. D for submersion trials at the Haustenbeck testing grounds (Sennelager troop training grounds). The bracket mounts for the side skirts turned out to be too weak and broke off at the weld joint. When the vehicle entered the water the skirts bowed out from the water resistance. The submersion trials, which began on 18 July 1942, showed that the vehicle developed numerous leak points.

In mid-June 1943 one Panther each was delivered for factory trials to the firms of Maybach in Friedrichshafen and the Süddeutsche Arguswerke in Karlsruhe.

Difficulties were also encountered with the key slot connection in the yoke of the torsion bar. Both torsion bars were joined together by a yoke on the side opposite that of the road wheel arm. Inside this yoke the two retaining cylinders were forced against a ring facet over a shim and a retaining brace. The open end rings and their connecting piece were designed too weakly and over time became loose, giving way to the pressure of the retaining shim. Once the connecting piece and the ring ends had been strengthened the problem with the shims becoming loose was corrected.

On 1 June and again on 15 June 1943 Generaloberst Heinz Guderian, in the capacity of his role as Generalinspekteur der Panzertruppen, paid a visit to both Panzerabteilung 51 and 52 at the Grafenwöhr training grounds. He examined his problem child, the Panther, the final drive and engine of which still displayed serious deficiencies. Of the roughly 200 Panther tanks already produced, only 65 had been accepted as technologically sound. In order to illustrate the time pressure under which the Panther was developed it should be pointed out that work on the assembly machinery (such as fixtures, etc.) began at almost the same time as they were being designed, i.e. there was very little time lag between design and implementation, and with the individual components already built and awaiting final assembly. All this at a time when a test vehicle had yet to be completed or had even begun trials. Roughly two months later, after the test vehicles had been finished, work began on production at a rapid pace. This meant that in the final assembly stage subassemblies which had been delivered from outside agencies were constantly having to be removed, modified, and reinstalled again. This with a work day of approximately 12 hours. The decision was reached, therefore, to modify the vehicles at the Grafenwöhr training facility. There the individual subassemblies were removed and modified as far as possible; it was, however, necessary to send some subassemblies to the manufacturer, where they were modified, shipped back to Grafenwöhr and reinstalled.

The Reichsbahn repair facility in Weiden was responsible for riveting the rims on the road wheels. MAN was required to completely redesign and manufacture the mountings for the optical sights. Under the direction of the labor teams, the soldiers of both Panzerabteilungen were heavily involved in the modification work.

On 16 June 1943 Guderian presented Hitler with his thoughts opposing the scheduled upcoming deployment of the new tank in Russia since the Panther, which due to time constraints had not been tested, was not yet ready for frontline operations. He was, however, overridden by the Reichsministerium für Rüstung und Kriegsproduktion — which had certified the tanks fit for combat — and Panther operations were therefore ordered by the Generalstab des Heeres (General Staff of the Army).

The Boehringer firm's hydraulic turret traverse was also built by the Schnellpressen-Fabrik Frankenthal beginning in April of 1943.

The following vehicles were assumed by the "Adolf Hitler Armor Program" in April 1943, set up to ensure that the troops were equipped with armored vehicles:
— Tanks without armament and accessories, but including special equipment for radio and optics according to specifications. with supplemental replacement parts
Priority Wehrmacht Contract Number 4911
— Sturmgeschütz and Panzerjäger vehicles, minus armament and accessories but including special equipment for radio and optics. with supplemental replacement parts

Priority Wehrmacht Contract Number 4912
— Armored 3-ton Zugkraftwagen (Sd.Kfz.251) including supplemental replacement parts.
Priority Wehrmacht Contract Number 4914
— Guns for above vehicles according to specifications Pak 40 and 43, with supplemental replacement parts

---

* On 11 May 1942 the Panzer-Panther-Kommission gave the green light for the construction and series production of the Panther combat tank. This act, unique in the history of tank development, led to serious problems shortly following the beginning of series production. The Panthers already assembled were taken apart at the Reichsbahn repair facility in Berlin-Falkensee, where the sub-assemblies were overhauled and re-installed. When new difficulties cropped up with the Panzerabteilungen 51 and 52 at the Grafenwöhr troop training range, these Panthers were overhauled a second time.

Priority Wehrmacht Contract Number 4915

A decision was reached on April 29th, 1941 that the Panther I would not only be built as a tank (Kampfwagen), but also as a Panzerjäger, or tank destroyer. The development of the Panther II, begun in February 1943, was continued. The use of skirts gave the side armor a significant improvement in shot penetration protection, so that even though the side plates on the Panther I were only 40 mm thick this was considered adequate protection. Feasibility studies were undertaken to determine the possibility of increasing the glacis armor to 100 mm; this would give a weight increase of roughly 0.5 tons.

A decisive factor in the decision to switch to the Panther II model was the practicality of using all-steel road wheels. If these could not be utilized on the Panther I a more rapid conversion to the Panther II would be necessary. The Heereswaffenamt/WaPrüf 6 accelerated the road test program for the all-steel wheels. The all-steel road wheels increased the vehicle's weight by approximately 2 metric tons. According to available calculations, with all-steel road wheels the Panther I would weigh nearly 46.5 tons and the Panther II 52.5 tons. Panther I Panzerjäger studies caused a minor delay in the production output at Daimler-Benz.

The importance of supplying the front-line troops with the Panther was manifest in the production program, which proscribed that from May 1943 onward a production rate of 250 Panthers per month was to be achieved.

Henschel announced on 3 May 1943 that by May 12th 50 Panthers would be delivered. Hitler determined on May 4th that no significant defects had turned up during road tests with the first two Panther test vehicles. As is so often the case, all problems first appeared when the troops began receiving the initial mass-produced vehicles. He recommended that in the future the practice of building individual vehicles for testing purposes be stopped, since there was the possibility that these might be manufactured with greater care and increased work effort — meaning that these vehicles could endure greater wear and tear than series-produced tanks.

In May of 1943 there was thought given to utilizing the Panther as a flame thrower. A single-wheel trailer was to be developed which would hold the flame fluid. Testing for this design was conducted with a Panzer III. A second type of "flame fluid" trailer with two wheels was shelved for the time being. The firm of Wegmann in Kassel was to research whether the Luftwaffe had a generator of roughly 30 kW available which would provide power to the electro-pump unit of the Schwade company. It was planned to mount the nozzles on the forward track sponsons.

In the Speer Ministry, Hauptdienstleiter Saur was forced to acknowledge that, instead of the 308 operational Panthers promised by 12 May, he could only report 100 vehicles ready at the most. He feared that Hitler might lose faith in the Panther program and fundamentally alter his previous disposition for operations. This would have to be prevented somehow, since in Saur's opinion the Panther was better than the Tiger. The conversion of 16 Panthers from the clutch-and-brake steering to the single-radius epicyclic steering required approximately 1000 man-hours per vehicle. MAH felt that it was not in a position to carry out this program alone.

On 15 May 1943 Henschel was instructed to produce the last 70 Panthers of its contract as Bergepanthers (armored recovery vehicles). Beginning with the 851st vehicle (Panther Ausf. A) it was planned to make use of a new type of turret.

Problems with the gun elevation drive were encountered as a result of a new requirement that the Panther also be able to drive cross-country with its barrel unlocked. The firm of Wegmann discovered a short-term solution which proved effective.

An exchange of letters took place on 28 May 1943 between the Zahnräderfabrik Augsburg (senior manager Johann Renk) and the Maschinenfabrik Augsburg-Nürnberg. It concerned the development of a hydrostatic-type steering unit* for the Panther. The Zahnradfabrik/Renk expressed its amazement that MAN was already in possession of all the basic knowledge necessary for the design of such a steering unit even before the beginning of discussions. At the first meeting MAN aired fundamental doubts about the function of this type of steering unit, and several times Renk was constrained to explain its operating characteristics. Since the first attempts to drive a filler pump from the engine r.p.m. were failures, Renk came up with the idea of installing an additional filler pump in the system. Renk wrote to MAN concerning this: "when we informed you of this solution, you expressed your amazement at the simplicity and effectiveness of this method we encountered. If these problems had been corrected earlier, you most certainly would have been inclined to share your knowledge and experience with us in order to reach our goal as quickly as possible."

---

* The first constant hydrostatic controlled differential steering was utilized for the steering drive in the years 1939-1940 in the French B2 tank.

Independent of testing initiated by MAN through the patent office, the firm of Renk claimed the right to free joint utilization within the scope of the patent petition Renk had already submitted.

In 1943 the Maschinenfabrik Augsburg-Nürnberg and the Zahnräderfabrik Augsburg eventually formed a partnership for developing and manufacturing hydrostatic steering units to be used in full-tracked vehicles over 9 tons total metric weight. The production facilities for manufacturing the steering units was to be set up in Marktredwitz. A hydrostatic steering mechanism was developed for the Panther which was fixed to a block along with the transmission. The power from the engine was transferred via the transmission to a bevel gear drive. A critical element of the shifting process was a hydrostatic oil drive with an axial-piston high pressure pump and a dual motor. Via a pitman shaft, this unit regulated the revolutions of the steering mechanism's planetary gearing, which in turn controlled the track speed. The drive power was relayed from the engine through the transmission to the hollow gears of the steering unit's two sun-and-planet gear drives. The planetary gears, which were driven by the inner-toothed hollow gears, rotated around the non-moving sun gears and set the planetary gear carrier in motion. Via the final drive, this in turn provided the power to the tracks and drove the vehicle.

When steering, the oil motor (in this case dual) drove the sun gears of the sun-and-planet drives in opposing directions. This caused one of the planet carriers to be driven faster, the other one slower. With the planet carriers one track ran faster and the other slower — thus steering the tank.

Since the oil motors' r.p.m. could be continuously changed when steering by regulating the oil pump, the vehicle was capable of performing infinitely variable turns.

With the hydrostatic steering unit, the following driving features were possible:
— straight driving, stable;
— infinitely variable turning radii in each of the transmission's gears;
— turning on the vehicle's axis

The axial-piston high pressure pump was driven by a shaft from the engine by the engine's r.p.m. If it were rotated to a certain angle, the axial piston stroke would be altered correspondingly and with it the pressure oil volume output. The secondary piece, in this case a dual motor, consisted of two axial-piston units fixed at a constant angle, to which the required pressure oil from the axial-piston high pressure pump was fed. The tank driver used a steering wheel to control the axial-piston pump's rotation along its cross axis. The amount of pressure oil fed depended on the pump angle. Since the motor's r.p.m. was proportional to the output volume, it could be continuously varied. The dual motor's direction of revolution could be reversed by changing the oil flow direction. Via a bevel gear pair the dual motor gave the neutral shaft a positive or negative revolution and, within a certain range, made possible an unlimited control of r.p.m. With its own r.p.m. the neutral shaft operated the

Continuous hydrostatic steering mechanism. Null shaft and twin motor can be seen in the foreground to the right.

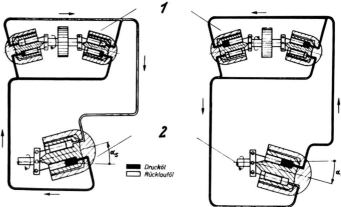

Oil flow in the axial piston high pressure pump and the twin motor. Left: coinciding rotational direction of input and output drive; Right: opposing rotational direction of input and output drive.
(1) twin motor (2) axial piston high pressure pump

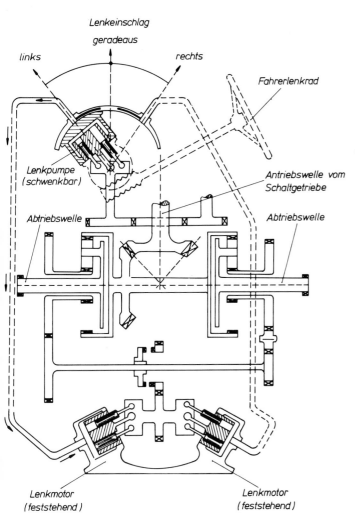

*Hydrostatisches Lenkgetriebe für Panther
System Renk
mit Thoma Hydraulik*

Principle of the Thoma hydraulics for the hydrostatic steering mechanism of the Panther.

Cutaway drawing of the continuous hydrostatic steering mechanism:
1 steering wheel, 2 axial piston high pressure pump, 3 feed line for the filler oil pump, 4 drive shaft for the axial piston high pressure pump, 5 engine input from engine r.p.m., 6 oil lines, 7 twin oil motor, 8 null shaft, 9 sun wheel gearing, 10 drive shaft from gearbox, 11 bevel gear, 12 planetary gearing, 13 planetary gearing, 14 output drive shaft to drive sprocket, 15 output drive shaft to drive sprocket.

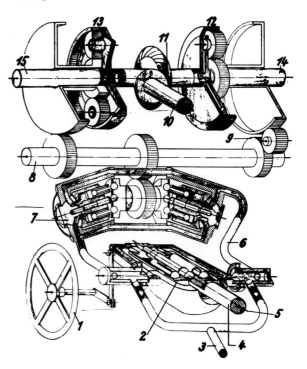

Cutaway photo of a continuous hydrostatic steering mechanism. The null shaft is visible, as is the twin motor, the axial piston high pressure pump in the lower portion and the steering wheel to the right.

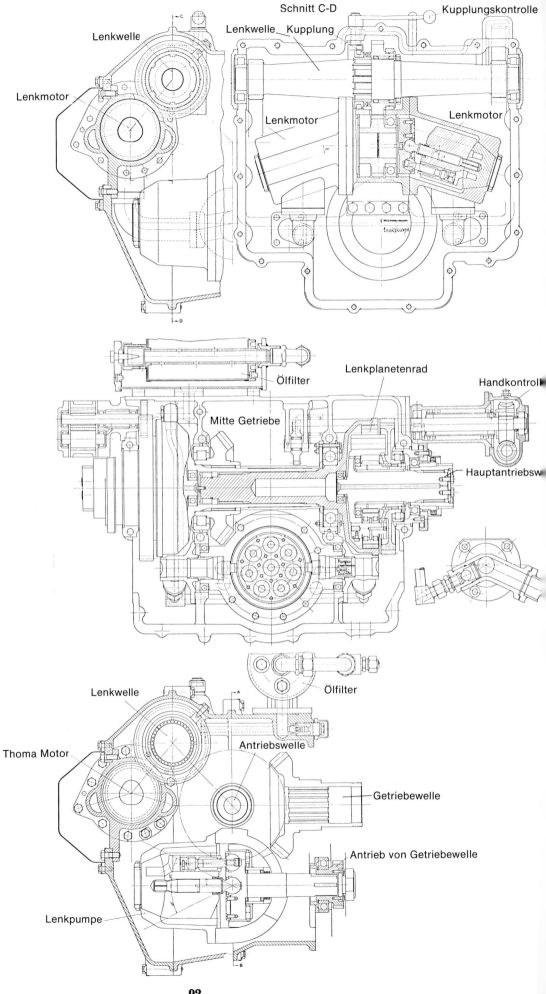

Additional details of the continuous hydrostatic steering unit of the firm of Renk.

sun gears of the left and right planetary drives in the steering mechanism against each other. In doing so, it provided the planet carriers with varying revolutions and enabled virtually unlimited turning radii with gears engaged in the transmission. The torque acting on the neutral shaft was also effective in the opposite direction of rotation. The same direction and speed of rotation of both planet carriers was achieved by blocking the neutral shaft from the opposing torque acting upon it.

This input was provided by the fixed deviation of the secondary part's two axial-piston units.

When driving straight the high pressure pump was put into neutral, yet in this position continued to function. The torque acting on the dual motor was checked by its axial pistons via the inactive oil shafts. Since both planet carriers therefore had the same direction and speed of rotation, the speed of both tracks was the same and the vehicle proceeded in a straight line. With the vehicle at a stop and the transmission in neutral, the steering wheel was used to rotate the axial high pressure pump driven by the engine r.p.m.; in turn the dual motor relayed — via the neutral shaft — an equal but opposing torque to the sun gears. Dependent upon the same dimensions of the planetary drives, the planet carriers transferred this correspondingly equal but opposing torque to the final drives. The speed of the outer track was therefore the same as that of the inner track — only in the opposite direction — and the tank could be rotated on its own axis. The design of the hydrostatic steering unit permitted a smooth transition from one driving state to another. Steering brakes and other similar arrangements, which had a relatively high amount of wear and tear and frequently required adjusting, often caused great difficulties in steering. With the hydrostatic steering mechanism, these were no longer necessary.

One of these units was installed in a Panther in August of 1944. During the subsequent thorough evaluation by the Verskraft in Kummersdorf a whistling tone was noted after a period of operation. This indicated that air was escaping from a pressure point in the pressure oil system. Shortly following the tone the vehicle lost its steering capability. The bleed-off pump would have to be given a greater capacity. The steering mechanism (MAN, Handradlenkung Argus, Teves and Daimler-Benz, anticipated manufacturers) was never put into full-scale production.

In August 1944 the design bureau of MAN was busy transferring to the Osterreichische Automobilfabrik AG in Vienna. There it would also undertake prototype manufacture as well as the construction of pre-production models.

In a written statement from 25 June 1943 to the OKH/Heereswaffenamt/WuG 6, Henschel declined the acceptance of Bergepanther production and pointed out that in the first half of August 1943 the initial Panther series (Ausf.D) would be terminated; the manufacture of the second series (Ausf. A) would have to be initiated during the course of the same month. According to MAN, this was to include various changes, including installing a ball mount for the radioman's machine gun, changing the arrangement in the fighting compartment and changing the ammunition storage.

In Operation "Zitadelle", which began on 5 July 1943 in the Kursk region, Panther Abteilung 51 and 52 (total of 192 tanks) were formed into a single Panther brigade and subordinated to the Army Group South. With its two Abteilungen, Panther Brigade 10 suffered serious losses in a minefield at the outset of operations. By the evening of the first day of operations, only 40 of the original 192 Panthers were still combat ready. Many others fell into enemy hands during the subsequent retreat after mechanical problems and lack of recovery vehicles rendered them immobile. Many Panthers were blown up by their own crew. The first Panther was destroyed by fire at the end of June while on its way from the Grafenwöhr troop training site to the train station there to be loaded. This was to occur frequently during operations.

Since it was found that the submersion seal installed in the turret bearing race caused the turret to become stuck, it was removed during the Kursk operation after the turret had been lifted off the superstructure.

The crews had hardly had time to become acquainted with their new equipment; due to the shortage of available time their training had been cut short. In many cases, too, there was a lack of the necessary combat experience. Thus it was that the use of factory fresh and untested Panthers in Operation Zitadelle (which ended on 13 July 1943) became a tragic chapter in the history of the German Panzertruppe.

During Panther operations in the Kursk region, Major Dipl.-Ing Icken from the Inspekteur der Panzertruppe in Berlin visited Panther Abteilung 51 and 52 and telegraphed to the Generalinspekteur der Panzertruppe the first unhappy news from the battlefield on the low number of Panthers still in operation. This report was

The Panthers of Panzer Abteilungen 51 and 52 seen being loaded in June 1943 for their first operations.

Most of the tanks used in the Panther's first operations had to be abandoned due to mechanical breakdowns and, since there was no adequate recovery equipment, were blown up by their own crews as they retreated.

The 18 ton half-tracks were only able to recover damaged vehicles in a train of three.

initially discounted by the Operations Department in the Führer's headquarters, then later was confirmed. On 10 July 1943 Guderian paid a call on the two previously mentioned Panther Abteilungen in the Army Group South area of operations near Bjelgorod and confirmed his fears regarding the Panther's lack of readiness for front-line operations. One must take into account the political climate which induced Guderian to send the following letter to Speer on 31 July 1943:

"Dear Reichsminister Speer,

As the most senior soldier in the Panzertruppe, I feel it is necessary for me to inform you of the great satisfaction of the Front with the new weapons following the first battle with the "new tanks", and to express our gratitude for these. The armor crews on the Front were particularly pleased with the unheard of performance of the advanced Panther gun; thanks to its superior performance it was repeatedly successful in destroying the T-34 (which for so long had proven difficult to engage) — even at 3000 meters.

When this awareness of the superiority during the course of the armor battle became widespread among the Panzerwaffe, it naturally formed into an appreciation by the soldiers for all those taking part in design and manufacture. I would therefore ask that you express in the broadest sense my thanks as the thanks of the Front to the entire industry involved. I am, with the expression of special allegiance

Heil Hitler
Respectfully Yours,
(signed) Guderian

On 28 September the first reports regarding the Panther operations at Kursk from 5-13 July 1943 became available. One Abteilung reported 25 engine failures within nine days, primarily due to piston rod bearing damage, water in the exhaust, damaged pistons, tears in the cylinder sleeves, burnt cylinder head gaskets and broken piston rods. On many occasions the engine became overheated. The engine performance in some vehicles dropped off after a short time. The engines had a high rate of oil consumption. Spark plugs became coated with oil. The danger of fire was very high, since lines and connections were often not sealed. The main clutch proved reliable as long as the vehicle was not utilized for towing. The transmission also functioned without problems. The final drive was too weak and had a high failure rate. The running gear proved to be completely successful; a few track tension adjusters became loose. The driver and radio operator hatches had a tendency to become stuck, so that the crew drove with open hatches even during operations. The hull superstructure above the driver and radioman was penetrated by shots deflecting off the lower part of the gun mantelet, necessitating a change in form. A reinforcement of the hull side walls was recommended, and the side skirts often were lost. The troops were enthusiastic about the new gun; no difficulties were encountered with it. The majority of enemy tanks were taken out at a distance of 1500 to 2000 meters. Turret ventilation could be improved somewhat. A gimbal mount was demanded for the radioman's machine gun. The optical mounts were too weak, the turret traverse too complicated. In the end, it was determined that the Panzerkampfwagen Panther had proven effective. Once the design errors were rectified the majority of technical problems were considered acceptable. In any case, the supply of replacement parts would have to be ensured. Recovery vehicles for the units were of critical importance.

On 13 July 1943 the Kraftfahrtversuchsstelle (Vehicle Testing Facility) of the Heereswaffenamt Prüfwesen in Kummersdorf-Schiessplatz relayed that a steering unit for smaller turning radii had been installed in a Panther, chassis number 210004, for test comparisons with production models. The calculated steering radii were to be as follows:

	Series production	Shortened radius
1st gear	4.8 m	2.5 m
2nd gear	11.0 m	6.3 m
3rd gear	18.0 m	10.9 m
4th gear	30.0 m	17.5 m
5th gear	43.0 m	26.2 m
6th gear	61.0 m	37.4 m
7th gear	80.0 m	50.5 m

Testing showed that it had worse drive handling on both roads and cross-country stretches than the production version. Since the standard brakes were not always adequate for steering, it was constantly necessary to either rapidly shift to a different radius or to bypass this step completely and simply engage the track brakes. On 16 August 1943 it was announced that the OKH Wa J Rü (WuG 6), in agreement with WaPrüf 6, had forwarded a recommendation to the chairman of the Hauptausschuss Panzerwagen und Zugmaschinen (Committee for Tanks and Halftracks), director Dr.-Ing. Rohland of the Vereinigte Stahlwerke, that the redesigned hull be introduced beginning with the 2801st Panther I vehicle. In August of 1943 the firm of Lanz in Mannheim was brought into the Panther program, providing for the assembly and welding of the hull body. The individual components, however, were not to be manufactured by Lanz, but by subcontractors in France. Since Lanz was only set up for the purely mechanical production, it was not given the task of producing the heating and sighting components.

Even as late as September of 1943 the bottlenecks in material supply had not yet been eliminated. On 15 September 1943 Henschel stated that from 9 September to 15 September 1943 it had not received a single Panther hull and therefore its drilling assembly line had not been in operation for days. For the most part, the requests of the troops were considered with the new Panther model, the Ausf. A, which would be introduced with the 851st vehicle in August 1943. The Panzerkampfwagen Panther, Ausführung A (Sd.Kfz.171) was first shown in September of 1943 and came equipped with a ball mount for the radioman's machine gun position developed by Daimler-Benz. Another external recognition feature was the monocular TZF 12a turret sight, which only required a single peephole opening in the gun mantelet. The main gun elevation was reduced to -8 deg. to +18 deg. (previously 20 deg.). The number of rounds carried remained at 79. The turret position indicator inside the turret now came

Generaloberst Heinz Guderian, the creator of the German Panzertruppe and its most senior soldier.

with a numerical dial plate with divisions from 1 to 12. The turret basket consisted of a central piece made of tubing and four side pieces which could be pulled up. By this time the Boehringer Sturm L 4 S oil drive was being installed; this ran off the engine r.p.m. and powered the turret traverse. The commander's cupola had already undergone fundamental changes during the Ausf. D production. It consisted of the cupola housing, the cover with arm, the periscope housing and the cupola ring. This latter piece served as a mount for the anti-aircraft MG 34 machine gun. A periscope was fixed into the right portion of the turret roof to provide an external view for the loader. According to information from OKH (CH H Rüst and BdE) from 25 January 1944, 76 g 16/17 Nr. 3022/44, the Panther chassis numbers were broken down as follows:

Panther, Ausf. D chassis no. 2100001-210254 211001-216000
MAN, from   210001   DB, from   211001
Henschel, from   212001
MNH, from   213001

Panther, Ausf. A
chassis no. 210255-211000   151000-160000

On October 2nd, 1943 the Maschinenfabrik Augsburg-Nürnberg retrospectively took a position on accusations of inadequate Panther manufacturing. According to MAN, the Reichsministerium für Rüstung und Kriegsproduktion had explicitly dispensed with prior testing and the usual pre-production series — errors of judgement which brought about all the initial problems in the series production. As the company which first began production, MAN alone was now basically responsible for overcoming these problems for itself and the companies license-building the Panther. Additionally, machinery and labor were constantly made available too late and in too little numbers. As the developing firm, MAN was repeatedly assigned special projects, among which were:

A late version of an Ausf. D Panther tank. The commander's cupola was changed, and the communications hatch in the turret side plate was done away with. The pistol ports were still used.

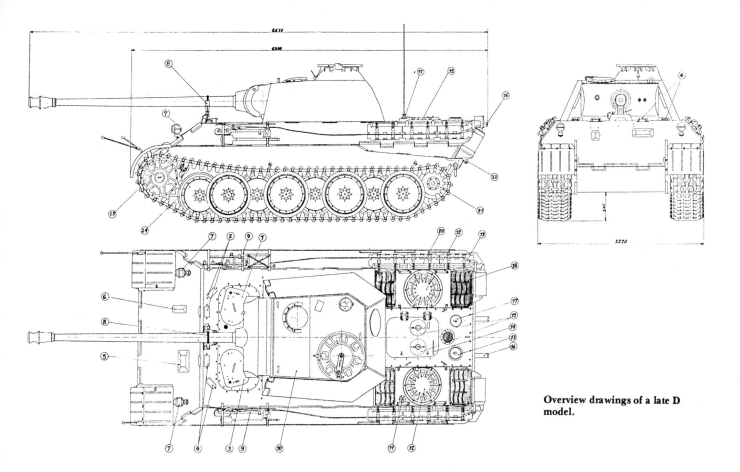

Overview drawings of a late D model.

Panther tanks during operations in Russia and Italy.

Chassis finishing: the road wheel arms have been assembled, as has the idler wheel arm.

The second series model of the Panther, Ausf. A, was produced in large numbers at three manufacturers.

Final drive with return roller has been installed.

Idler wheel in its mounted position.

Panzerkampfwagen Panther, Ausf. D (late model) (Sd.Kfz.171)

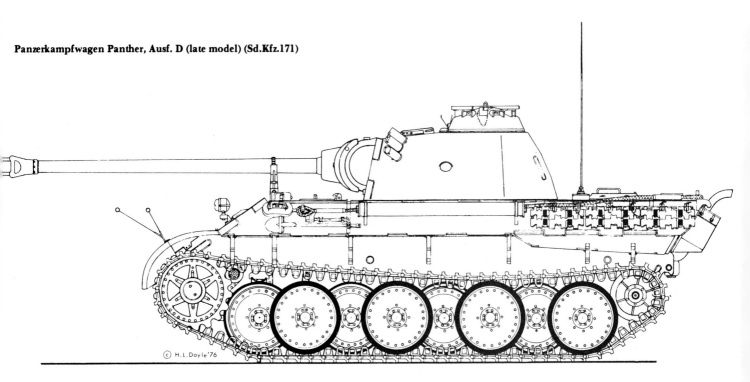

Panzerkampfwagen Panther, Ausf. A (Sd.Kfz.171)

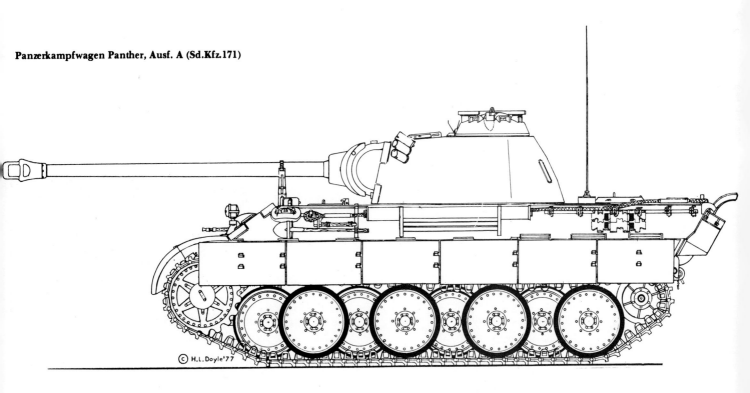

The road wheels have been mounted.

With the exception of the tracks, the running gear is now complete.

The hull from the front shows the radio operator's ball mounting for the MG 34. The Zimmerit coating for protection against hollow charges has been applied.

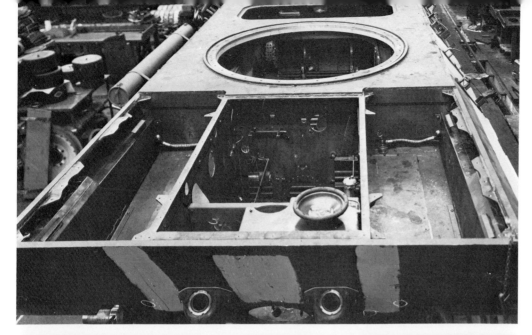

A rear-view of the hull prior to the installation of the engine and cooling system.

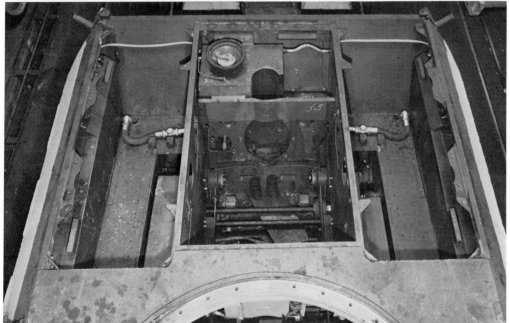

Engine compartment before installing the components. The fuel tanks have already been assembled.

The cooling system is installed.

— modification program both at the Berlin-Falkensee Reichsbahn Repair Facility and at the main facility in Nuremberg. A total of 90 Panthers and 103 single-radius epicyclic steering mechanisms were converted at the Nuremberg plant. 17 modified steering units were handed over to the Reichsbahn Repair Facility in Berlin-Falkensee.

— modification program for Panzer Abteilung 51 and 52 at the Grafenwöhr training area and at the Panther training course in Erlangen. 563 Panthers were modified under the direction of MAN.

— initial problems with the manufacture of single-radius steering mechanisms. Henschel had been building these since August 1942, while MAN joined in production beginning in February/March 1943. Through comprehensive testing MAN succeeded in alleviating the considerable difficulties which remained.

— setup of a plant for manufacturing replacement parts in Leipzig. In place of the destroyed Kongresshalle, the Papiermaschinenfabrik K. Krause, Leipzig, was made available for this purpose after lengthy negotiations.

A four-view of a Maybach HL 230 engine ready for installation.

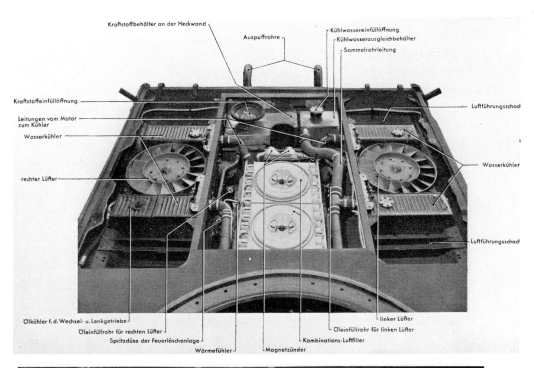

Engine compartment with components installed.

Covered engine compartment with protective gratings for the cooling air.

View into the forward compartment of the hull shows the foot pedals for the driver on the left. On both the left and right the disc brakes can be seen. A shock absorber is seen behind the brake on the right. The ball mounting for the machine gun has been installed in the radioman's position.

MAN supplied a total of 730,208 pieces having a weight of approx. 476 metric tons for these various modification programs. 183,000 assembly hours were spent on the modification programs, which equates to 130 skilled laborers working continuously for five months. In spite of the startup difficulties and the first air raids, the following deliveries had been made by 31 July 1943:

	Quota	Actual	Difference
MAN	224	209	15
Daimler-Benz	221	202	19
MNH	196	184	12
Henschel	115	116	-
	756	711	46

The state of deliveries for MAN on 30 September 1943 was affected by two massive air raids in August of that year, where 52,000 work hours were lost in tank manufacturing alone. In addition a delay of nearly 14 days was encountered in turret production due to a premature introduction of a new turret design on the Panther Ausf. A. Of the Panthers delivered up to 30 September 1943, 61 were Befehlswagen (command vehicles); these demanded a significantly greater amount of work and were only built by MAN. On 15 October 1943 MAN announced that it intended to negotiate with companies in Switzerland for the manufacture of cast armor parts. Additional modification programs were set up in the Panzer-Heereszeugamt

Transmission prior to installation.

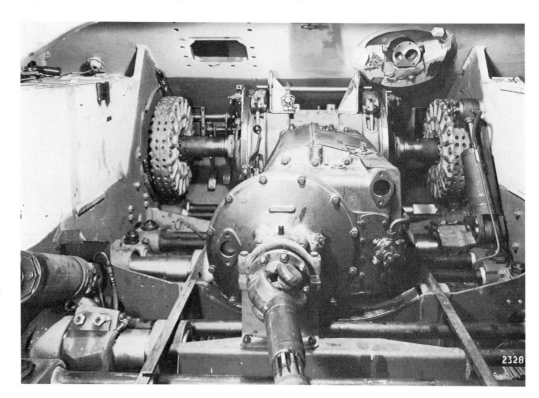

The installed transmission. A shock absorber can be made out in the left foreground. A drive shaft connects the engine and transmission.

Explanation of the functions in the forward hull.

Details of the driver's and radio operator's stations.

Driver's (left) and radioman's stations with seats.

Magdeburg-Königsborn. By 15 November 1943 150 vehicles were to be modified there, but 170 Panthers had already been taken in by the 8th of that month. This modification program was basically completed by the end of November 1943. A regrettable decision was made to send 60 Panthers to the Heeresgruppe Nord on the Eastern Front where they were to be buried as part of a fortification system. This undertaking was not carried out after Guderian, in his capacity as Generalinspekteur der Panzertruppen for Hitler, had expressed his objections. The occasion for this venture was a complaint by Hitler

The radio operator's station prior to installing the seat; the box for the radio equipment has not yet been built in.

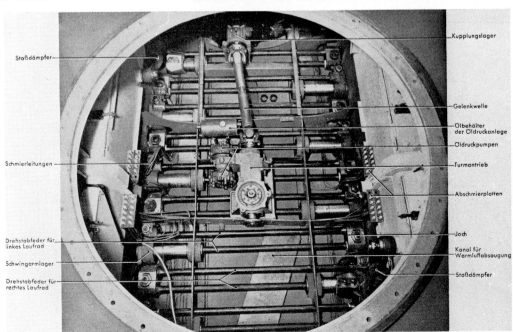

Power train and suspension.

A view into the fighting compartment shows the turret traverse drive in the middle. In the upper portion, beneath the turret floor, the batteries can be seen. Some of the ammunition holders are already in the turret basket.

View into the fighting compartment from the opposite direction.

Bulkhead between the fighting and engine compartments.

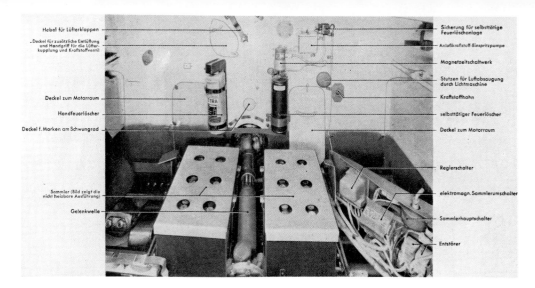

Left: Additional details of the firewall.

Below left: Main fuse box in the chassis.

Below: View from below showing the mounting of the main weapon.

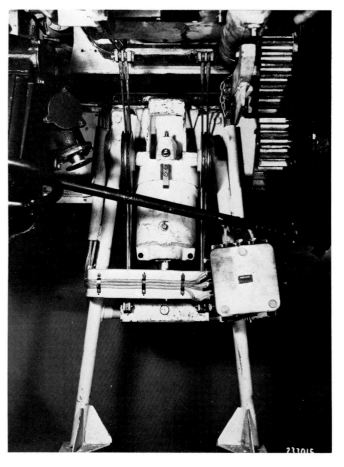

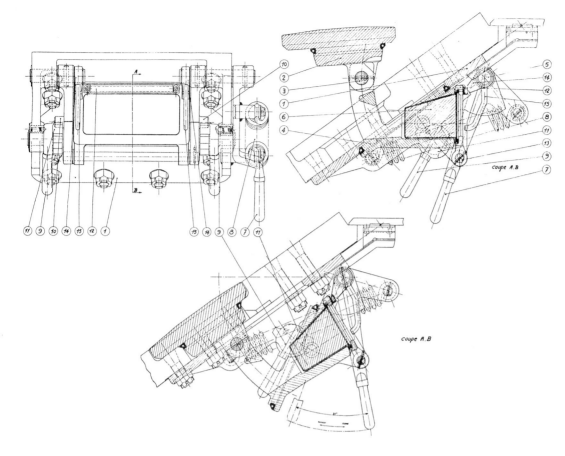

Details of the driver's hatch.

Walkaround view of the driver's station with entry hatch.

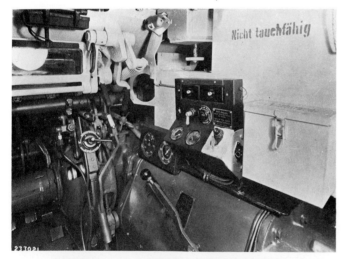 

 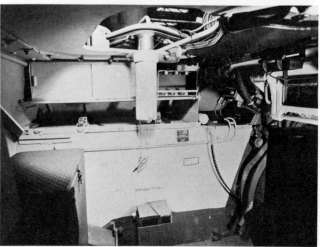

Details of the fighting compartment showing ammunition racks and other equipment items.

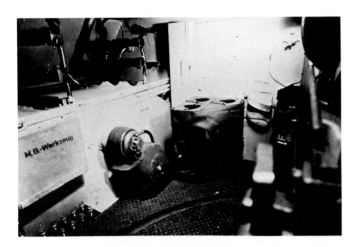

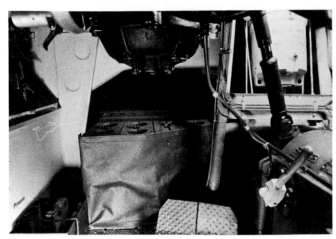

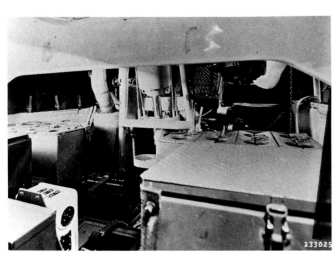
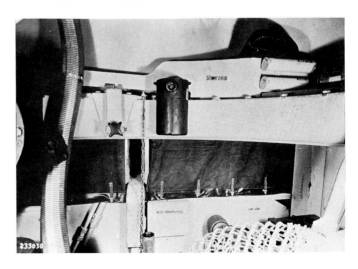

Above right: After having been delivered in completed form the turret is set onto the hull.

Center right: View showing the breech of the gun, the radio operator's machine gun to the right of the gun and the gunsight

Below right: Finishing the interior of the turret.

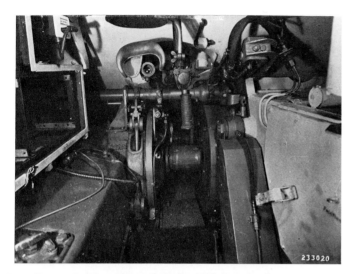

Left: Walkaround of the radioman's station with entry hatch.

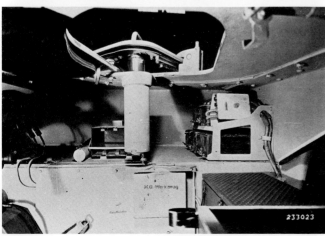

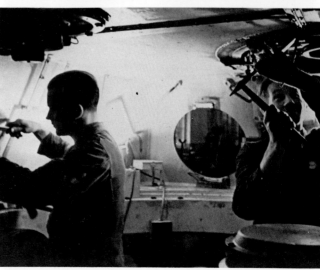

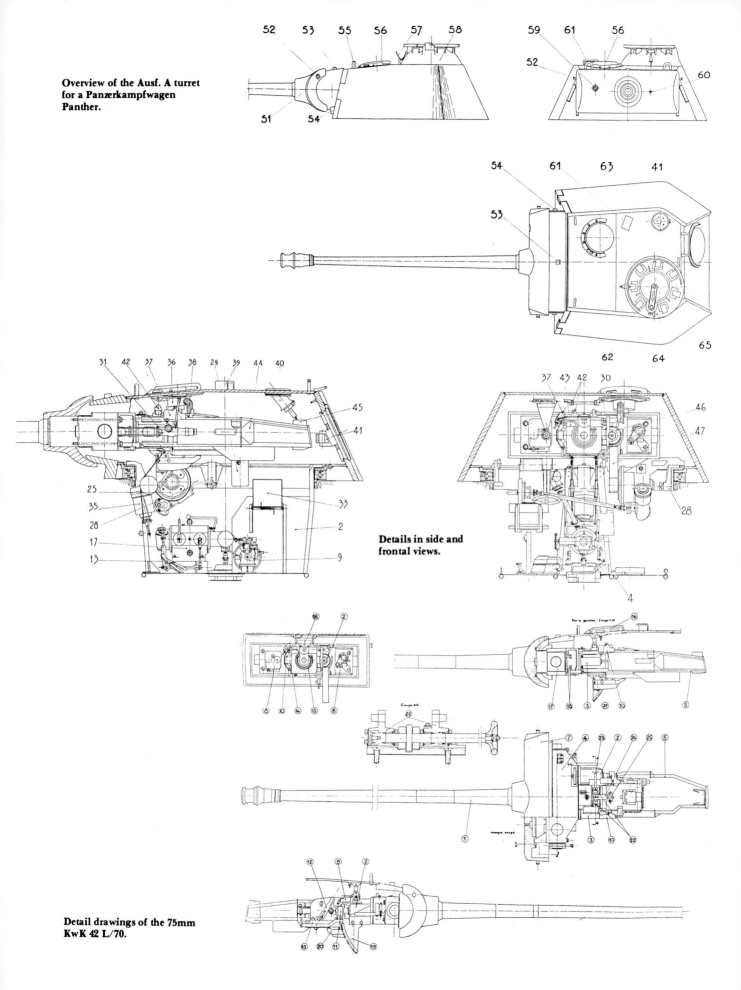

Overview of the Ausf. A turret for a Panzerkampfwagen Panther.

Details in side and frontal views.

Detail drawings of the 75mm KwK 42 L/70.

Right and left side views of an Ausf. A turret with turret basket.

Frontal view of the turret with AA gun mount.

concerning the high Panther losses encountered on the Eastern Front. Hitler noted: "If we had taken these Panthers we had lost on the Eastern Front — of which there were no less than 600 — and fixed them solidly into the defense line, we would then have had an absolutely foolproof armor protection. We have nothing like this." Hitler once again addressed the Panther engine on 17 December 1943. He was convinced that there was no possibility of switching over to a different engine at that time. Nevertheless, all resources should be devoted to the matter of developing an air-cooled diesel engine. Around this time Daimler-Benz in Berlin-Marienfelde once again made an attempt to introduce its MB 507* liquid-cooled diesel engine, without supercharger, into the Panther program. This was countered by the fact that, in a meeting

---

* Water-cooled 4 stroke diesel engine, 12-cylinder in V form 150x180 bore/stroke. 42.3 liter swept volume. 850 hp at 2300 1/min.

Details of the turret interior features below and above the gun. (For English translations of features see pp. 181-183)

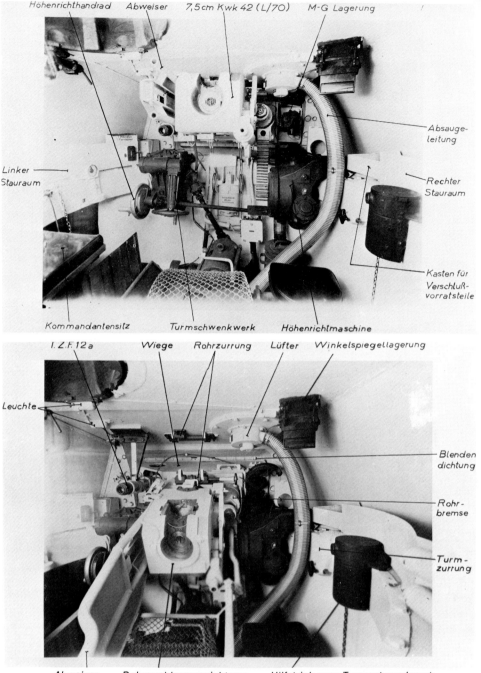

on December 21, 1943, Speer had expressed his complete confidence in Dr.-Ing. e.h. Maybach that all teething troubles with the HL 230 engine would shortly be eliminated and then would be able to meet all requirements. The creation of an alternate solution was not discussed during the meeting. Additionally, the suitability of the Daimler-Benz diesel engine in a tank had not been proven. A total of 108 Panthers were tested and accepted from the road test facility at MAN in December 1943. These had been driven 13,250 km. A total of 1783 Panthers had been produced by all manufacturers in 1943.

During a meeting of the Arbeitskreis Panzer (Tank Work Group) on 4 January 1944 there was discussion about consolidating the Panther and Tiger tank programs. The decision of the Panzerkommission (under the direction of Dr.-Ing. Stieler von Heidekamp) indicated that the Panther II and Tiger II would be developed to the design stage and that the Tiger II would be put into production. It would be necessary for a new hull to be designed for the Panther II prior to it being introduced into production. The main obstacles to a consolidation were:

— opposing direction of rotation of the steering differential drive shafts and
— different axle spacing of the final drive.

Left side of turret basket.

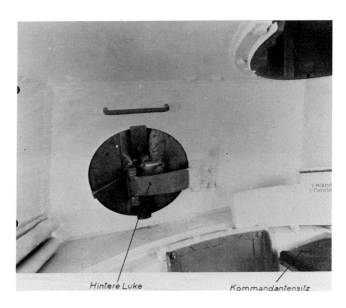

Turret roof with ventilator and periscope mounting for the loader.

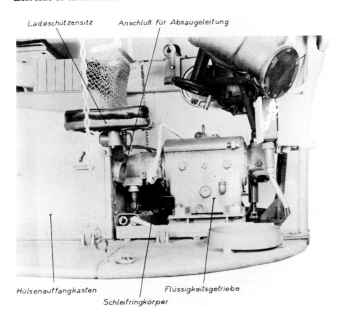

Right side of the turret basket.

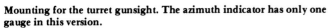

Rear hatch, portion of the commander's seat and cupola.

Mounting for the turret gunsight. The azimuth indicator has only one gauge in this version.

Junction boxes for the electrical connections inside the turret.

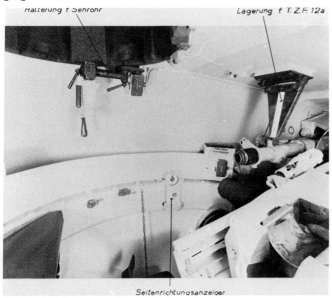

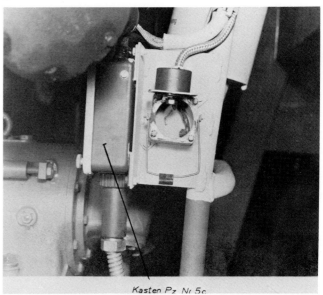

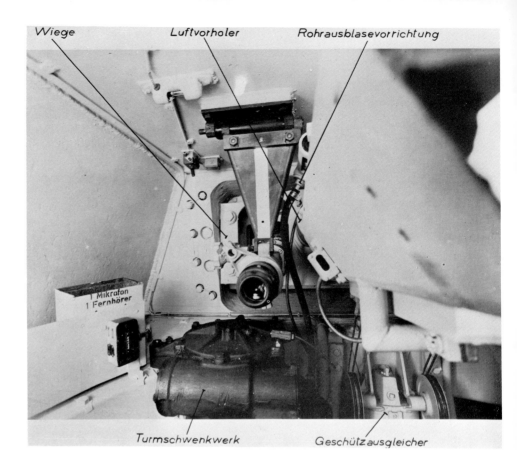

Gunner's station within the turret.

In adapting the Panther II to the Tiger II it would be necessary to install the steering unit and the final drive of the Tiger II, steps which would involve a comprehensive design modification to the Panther's hull. If a drop in production were taken into account as a result of the consolidation, the following separate components could be used for both the Panther II and the Tiger II: HL 230 engine; cooling assembly; transmission (either AK 7-200 or OLVAR B); Tiger II steering unit, Tiger II final drive; all-steel road wheels. The non-common parts would be: road wheel hubs; swing arms and mounting brackets; torsion bars. Required changes to the Panther Ausf. G were therefore needed with the hull, steering unit, final drive and running gear. Measures were undertaken to strengthen the Panther's final drive in conjunction with the consolidation. For reasons of durability, it was no longer possible to avoid installing a final drive with planetary gearing. If the Panther Ausf. G were retained there were two avenues:
— a final drive based on a recommendation by MAN which made use of Tiger parts, or
— a final drive using components of an all-new design.

If the Panther II were put into production the final drive of the Tiger II would be used. Since the MAN final drive could only make use of the center hollow gear wheel, the carrier and the planetary gear wheels, it was recommended that this design be constructed using new parts.

The work time involved in building a planetary-gear final drive was 2.2 times greater than the machining involved in the previous final drive.

It was the decision of the Panzer commission not to press on with the merger of the Panther and Tiger at that time, since the anticipated drop in manufacturing output was not felt to be tolerable.

Nevertheless, work was to proceed on the MAN final drive using new components. The necessary machinery

Barrel travel lock and lifting mechanism for the commander's hatch.

118

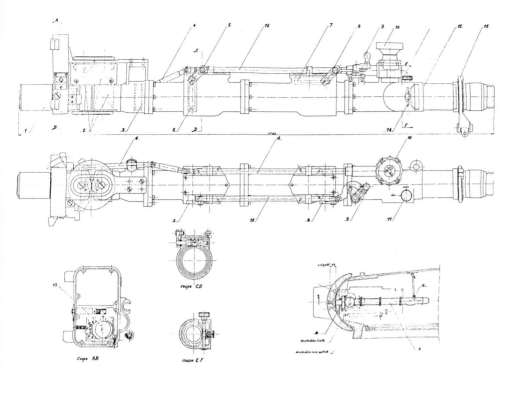

Details of the TZF 12a turret gunsight.

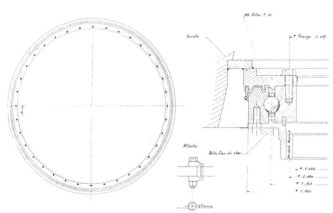

Turret race ring and bearing race.

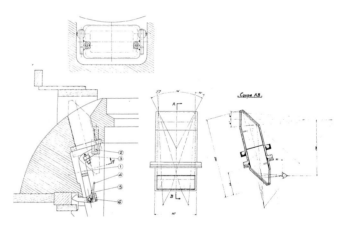

Periscope and mounting for the commander's hatch.

Details of the rear turret hatch.

Details of the turret housing.

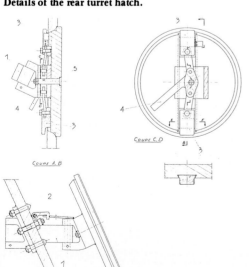

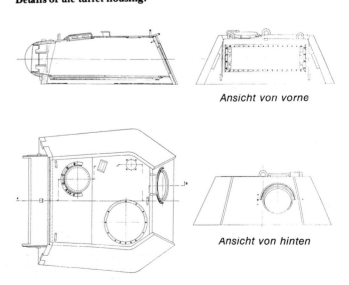

Ansicht von vorne

Ansicht von hinten

equipment was to be made available, and the consolidation was to proceed.

The double-row toothed tracks used on the Panther had proven inadequate due to the tendency of the teeth to break and were too weak for an all-steel running gear. A strengthened single-row track was recommended. This track was comprised of cast metal links, each having a single tooth and two grousers. These links were separated by spacer links of rolled steel which did not have the guide teeth. The introduction of this track mandated an enlargement of the track adjusting shaft and a modification of the rear hull wall. The all-steel running gear was planned to be interleaved. Both the track tensioning design and the all-steel running gear were scheduled for introduction in August of 1944. The speed in reverse gear was expected to be increased with the AK 7-200 transmission. ZF produced test pieces for replacing the gears which would raise the speed from 4 to 5.7 km/h.

Daimler-Benz reported to Reichsminister Speer on 2 February 1944 that an additional 10,000 m2 of production facilities and 420 more workers would be made available for Panther production.

In mid-January 1944 the Forschungskreis "Suspension and Flotation of Tracked Vehicles" was founded in Nuremberg. Dr.-Ing. habil. Ernst Lehr, as the best expert in this field, was called as its director. The Forschungskreis "Suspension and Flotation", which was already in existence under the direction of Dr.-Ing habil. H. Maruhn, would now devote itself primarily to suspension matters

A Panther tank delivered to Sweden in 1943.

of wheeled vehicles. It was also given the special project of the suspension on the Maultier half-track.

There was interesting discussion even during the first meeting of the Forschungskreis. Oberst Holzhäuer, Office Chief of WaPrüf 6 of the Heereswaffenamt, felt that the time had arrived when even the heaviest tanks could no longer offer adequate protection. It was therefore incumbent that the speed and maneuverability of the tank be emphasized. The goal of firing from a moving tank was yet to be attained. This could only be achieved with the introduction of a gyroscopic stabilized optical system. A stabilization of the weapon was also not yet possible, due to the strength required in the gun design. At this time there were tests being conducted with the gyroscopic optics which had already led to some success. Oberstleutnant Dipl.-Ing. Stollberg, Heereswaffenamt WaPrüf 6, indicated that the production costs for vehicle construction were high because of the double torsion bar

Panzerkampfwagen Panther, Ausführung A, from above.

Left side view of the Panther Ausf. A, minus and with side skirts.

suspension. The swing arms and their mounting were particularly troublesome. Oberst Dipl.-Ing. Esser, director of the test facility for Panzer und Motorisierung in Kummersdorf-Schiessplatz, replied that with the double torsion bars it was necessary to make use of the entire hull floor, which placed limitations on the designer with regards to accessibility to the engine and other critical vehicle components. Chief engineer Maennig of Daimler-Benz described the torsion suspension as a highly refined type of suspension and urged that the quality of leaf springs be increased accordingly. Dr.-Ing. Lehr replied that leaf springs could not meet the same demands as torsion bars. They could not, for example, be used on a vehicle designed for rapid speeds. In any case, up to that time only with the aid of torsion bars was it possible to design a suspension system for the Panther which ensured a travel stroke of 510 mm (measured at the road wheels), a pitch rate of 30 undulations per minute and a fatigue life of more than 10,000 km.

On 27 January 1944 Hitler was informed of the percentage of monthly losses of the individual tank and Sturmgeschütz types. In particular, these figures showed a growing increase in the number of operational Panthers since October 1943 and a drop in the number of Panthers lost in operations. Speer took this occasion to report on the experiments conducted by MAN using incendiary charges on the Panther to feign damage; these experiments were to be continued at an accelerated pace.

A protocol from 2 February 1944 referred to a discussion between Dr.-Ing. e.h. Maier of the firm of ZF and Dr. Linnebach of Heereswaffenamt/WaPrüf 6 concerning the use of the steering mechanism as a type of transmission by skipping a gear through simultaneously activating the auxiliary brakes on either side and the steering linkage (steering lever).

In order to reduce the road speed with the single-radius steering unit at constant engine r.p.m. it was necessary to release both auxiliary brakes and close both steering clutches.

The following units were considered for the proposal:
— MAN single-radius epicyclic steering mechanism
— MAN/ZF double-radius epicyclic steering mechanism
— MAN/ZF double-radius electro-steering mechanism

A speed decrease of 1.75 km/h was possible with the single-radius steering unit when a reduction of 1:1.28 (in second gear) was engaged. This transfer ratio applied equally for all gears. With the double-radius unit a change of 3.06 km/h was possible when engaging the smaller radius with a reduction of 1:1.61 (in second gear). With a 7-gear transmission, this transfer ratio dispensed with the first gear. The speeds were 5.03 km/h for the double-radius electro-steering mechanism in conjunction with the AK 7-200 transmission in the steering drive gear. The speed in first gear of the production AK 7-200 was 4.01 km/h.

In comparison, the steering drive gear of MAN's single-radius steering mechanism, with 6.34 km/h, was simply too great. A change in the second gear of the AK 7-200 transmission made an assimilation to the previous speeds possible. The following data was compiled:

Rear side view of a Panzerkampfwagen Panther Ausf. A.

Front view of a Panzerkampfwagen Panther Ausf. A.

Right side view of the Panther Ausf. A, minus and with side skirts.

Comparison of the A model both minus and with side skirts. The vehicle on the right also has an anti-aircraft machine gun mounted on the commander's cupola.

These rear end photos of a Panzerkampfwagen Panther Ausf. A show the improved exhaust cooling in the right photograph and the different method of carrying the jack so as to improve accessibility to the maintenance openings.

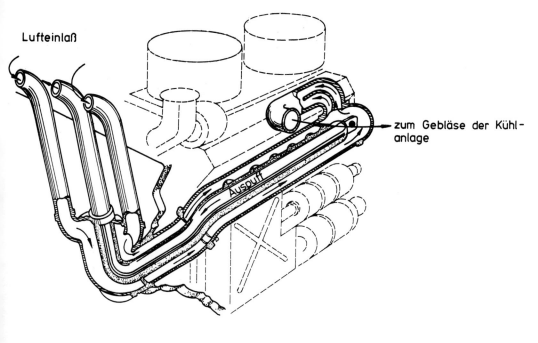

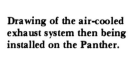

Drawing of the air-cooled exhaust system then being installed on the Panther.

Steering unit	MAN	MAN/ZF	MAN/ZF	
Radii	1	2	2 electro-steering unit LE-185	
Gearbox	AK 7-200	AK 7-200	12 E-185 G electro-gearbox	
Maximum speed steering gear	large radius 6.34	small radius 6.82	large radius 5.03	small radius 6.63
2nd gear		8.09	11.42	
3rd gear		12.85	18.70	
4th gear		20.10	27.15	
5th gear		29.00	37.90	
6th gear		41.00	55.00	
7th gear		53.60	–	

Producing a Panzerkampfwagen Panther Ausführung G. The circular opening for the driver's scope which had replaced the glacis plate vision port can be seen on the far left.

A mating of the ZF electro-transmission with the electro-steering mechanism meant that the cross-country gear group could be dispensed with. The first gear, as the steering drive gear, reached 6.63 km/h in the large radius and 4.84 km/h in the small radius.

It was recognized from these determinations that a first gear could effectively not be attained using the single-radius steering mechanism in conjunction with the steering drive gear. With the small radius of the double-radius steering mechanism there was a greater potential of engaging first gear via the steering unit.

A Führer directive dated 27 February 1944 established that the vehicle would simply be designated "Panther", meaning that the earlier designation of Panzerkampfwagen V was dropped.

In March of 1944 those firms still participating in the Panther program — Daimler-Benz, Maschinenfabrik Augsburg-Nürnberg (97th March-vehicle) and Maschinenfabrik Niedersachsen-Hannover — converted over to the Panther Ausführung G (Sd.Kfz.171) with chassis numbers beginning at 121301, 124301 and 214001, respectively.

It was determined that the Panther I would continue to be built over an extended period of time, while for the time being the Panther II would not be put into mass production. The manufacturing simplifications, which had previously been incorporated into the hull of Panther II as a result of technical experience, would now be transferred to the hull design of the Panther I. On 11 May 1943 the

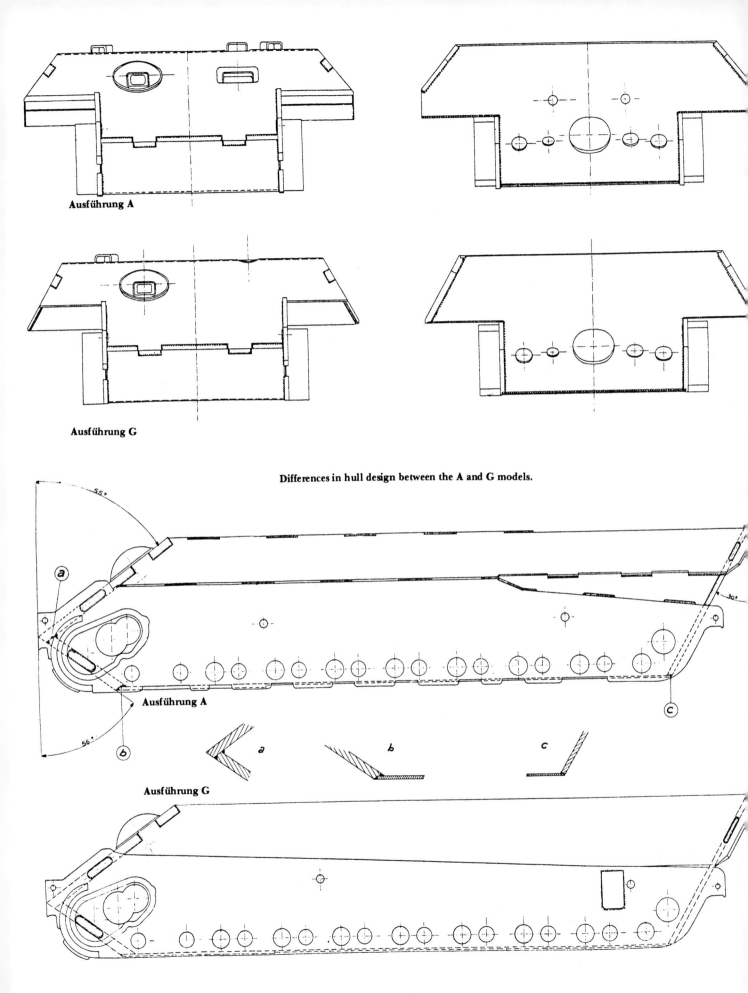

Differences in hull design between the A and G models.

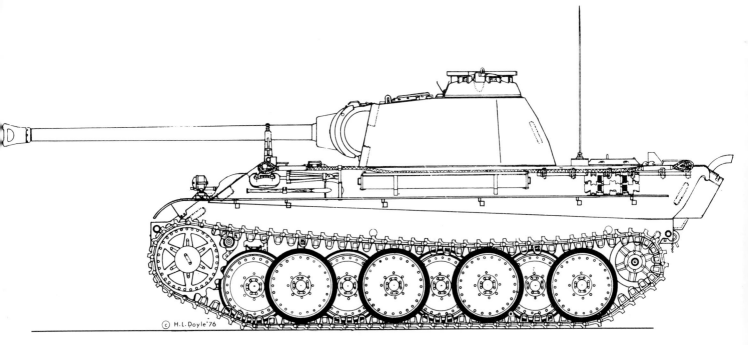

Panzerkampfwagen Panther, Ausf. G (Sd.Kfz.171)

Front and rear views of the Panzerkampfwagen Panther Ausf. G.

Driver's and radio operator's entry hatches, modified beginning on the G model. Here in closed position (radio operator's on the left, driver's on the right)

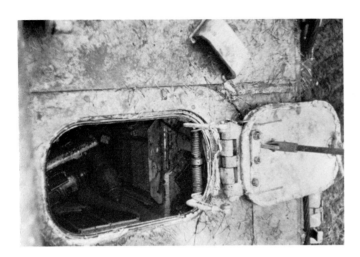

Entry hatches, open.

following manufacturing simplifications were relayed to the tank companies:

— The "kink" in the rear portion of the sponson and the angle in the sponson's bottom plate were eliminated accordingly, so that the bottom plate of the sponson sloped in a straight line from the point of the radiator rear edge to the forward edge of the sponson. The upper side wall was given an angle of 61 degrees (previously 50 degrees), and was increased in thickness from 40 to 50 mm in order to provide the same amount of armor protection. This resulted in a weight increase of roughly 305 kg. In order to offset this weight and to bring about a standardization/reduction in armor plate thicknesses it was proposed to decrease the lower front plate from 60 to 50 mm — giving a savings of 150 kg. For these reasons the forward hull bottom armor would also be produced with a thickness of 25 instead of 30 mm (as was planned on the Panther II), which gave an additional weight savings of 100 kg. A total weight reduction of 250 kg was achieved by the standardization of plate thicknesses, which meant that the strengthened side walls resulted in an overall weight increase of 55 kg.

— The bottom plate of the sponson was now 50 mm closer to the track than on the Panther II. MAN was concerned that the track would strike against the bottom plate when shifting and travelling across uneven terrain. The sponson bottom plate was to have no weld seams in the area affected by the track. The skirts were to be mounted in such a manner so that the planned anchor hooks would be welded to lower portion of the superstructure side wall. The Panther Ausf. G only made use of armor plating of 16, 25, 40, 50 and 80 mm gauge.

— The AHA/In 6 (Inspektion der Panzertruppen) demanded a center coupling on the rear of the Panther.

— For mounting the moveable driver's periscope MAN recommended producing the roof at this point with a thickness of 40 to 50 mm in order to provide adequate armor protection and a stronger base for the periscope. It was determined to thicken the forward 300 mm of the hull roof to 30 mm.

— All production simplifications which were to be undertaken on the Panther II were carried over by MAN to the modification of the Panther I (Ausf. G). The exhaust pipes were now equipped with flame suppressors, the

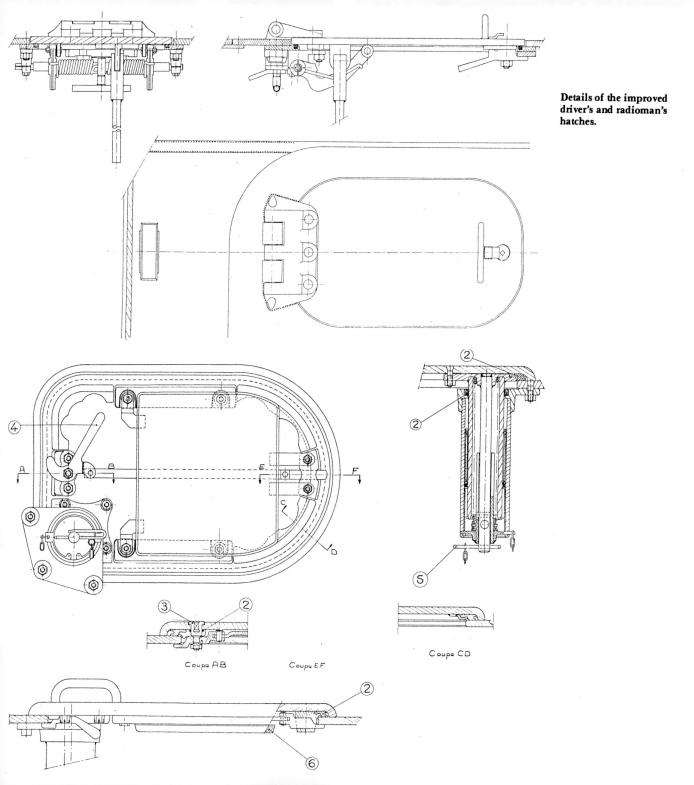

Details of the improved driver's and radioman's hatches.

Coupe AB Coupe EF Coupe CD

The cover for the fresh air intake was situated in the front between the driver's and radio operator's hatches.

Driver's pivoting periscope mounted in the hull roof.

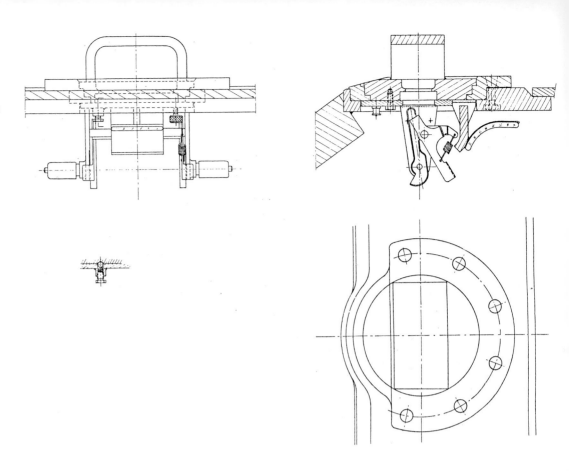

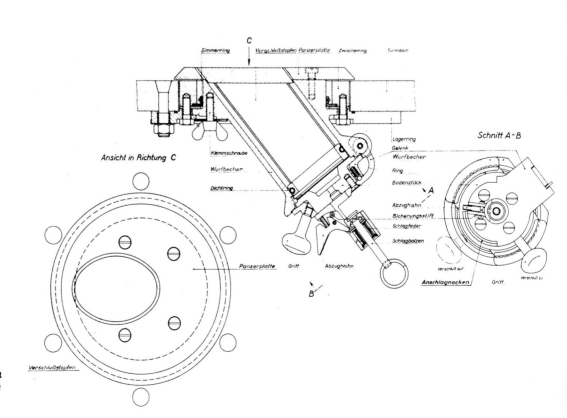

A close-combat defense weapon, similar to those installed in other combat vehicles, was fixed in the turret roof.

130

Panther G during a demonstration of army equipment. In the background is a Hummel ammunition carrier with the new resilient all-steel running gear of the Krupp firm.

This side view of a G model shows the downward angled edge of the tank hull above the running gear.

During operations different models of the Panther were to be found within a single unit. The front tank is an A model, while the rear vehicle is a G.

G model Panthers at the Maschinenfabrik Augsburg-Nürnberg prior to being loaded. It was necessary to camouflage the tanks in the factory in order not to make them such tempting targets for Allied fighter-bombers.

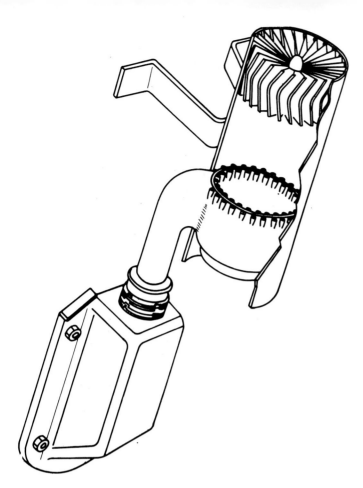

During the production of the G model a modified exhaust system was introduced with flame suppressors. The photograph and sketch show the arrangement of the system.

Panzerkampfwagen Panther Ausführung G during operations. The vehicle next to it is a Mittlerer Schützenpanzerwagen (Sd.Kfz.251).

idler wheel was changed and an improved driver seat was installed which could be raised and had extensions for all control levers. An improved heating system was planned for the fighting compartment.

The driver's vision port was dropped on the G series in favor of a rotating periscope (KFF) which was installed vertically. The eyepiece and the objective lens were both horizontal pieces; as with the turret scope a ray convergence design (entrance pupil) was advanced. The first edition proposed by the firm of Askania had a simple (1x) magnification at a field of view of 70 degrees. It had a considerable loss of light, in some conditions up to 85%. In spite of a large number of these optical devices being manufactured, they were not introduced because of the poor picture quality — the magnification at the edge of the field of view was greater that 1, meaning that the view did not remain fixed with the movement of the vehicle. Tests with the driver's periscopes of the firm of Meyer (Görlitz) gave better results, although the field of view was reduced to 50 degrees. By using photo-optical components and simpler prisms the picture quality and light throughout were improved and these devices grew in popularity. The final series model was therefore equipped with this device.

The driver's and radio operator's hatches were now mounted from the outside, thereby preventing the jamming encountered previously.

During the production run of the G series the gun mantelet was also modified. The mantelet's lower part was made thicker and vertical so as to eliminate the shot trap which formed below the mantelet. In this manner shots were prevented from being deflected through the hull roof.

It was now possible to carry 82 rounds for the main gun in the vehicle, along with 4800 rounds for the machine guns. The resilient steel road wheels were introduced beginning with chassis number 121052 due to raw material constraints. In addition, the tube for the barrel cleaning equipment was transferred from the left sponson and

During the production of the G series the turret gun mantelet was equipped with a deflector "lip", designed to prevent shells from entering through the hull roof.

Four-view of the **Panzerkampfwagen Panther Ausführung G (Sd.Kfz.171).**

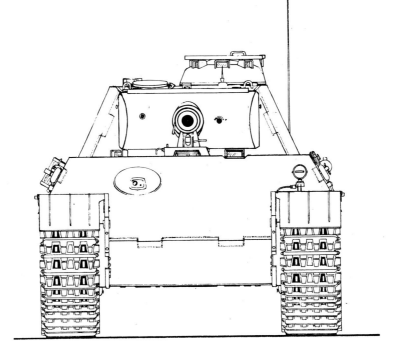

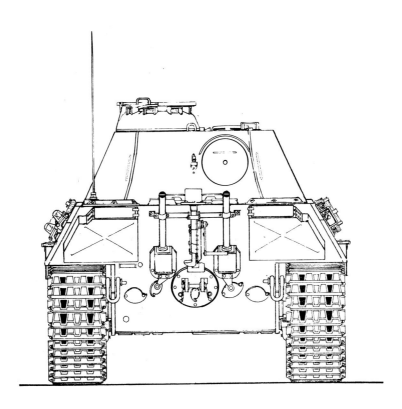

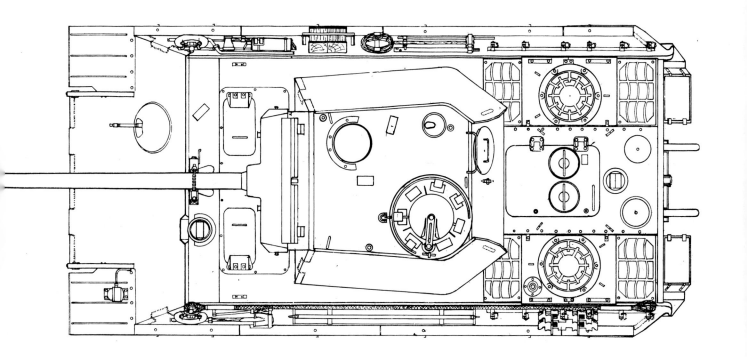
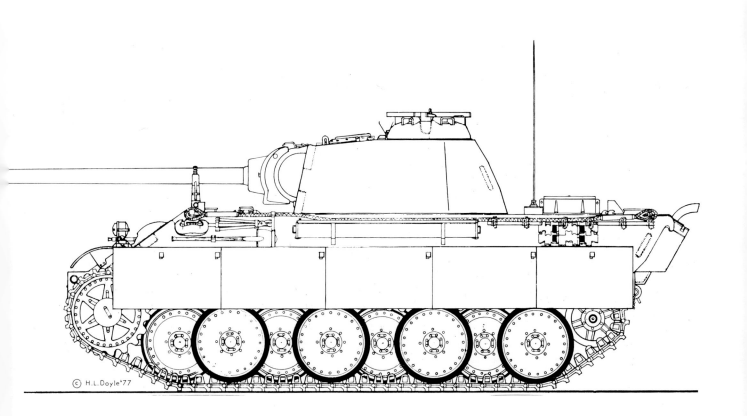

Close-up photo of the thickened gun mantelet on the Panther G.

Difference between the earlier mantelet and that of the G model. Sealing and trunnion pin are shown in the drawing. (F. Gruber)

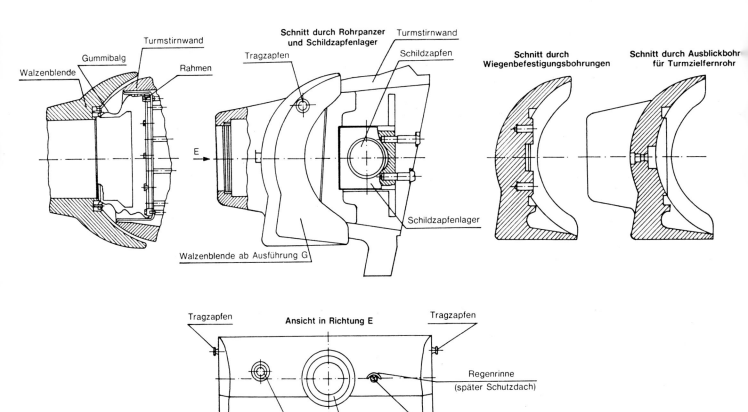

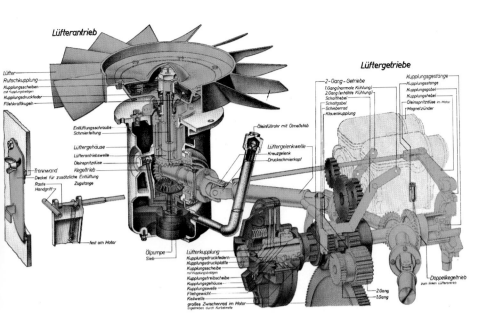

Fan drive for the cooling system, with two fan speeds for cooling the engine in both summer and winter.

The hydraulic pumps, driven by the main engine, used to operate the clutches and brakes of the steering mechanism. The bevel gear drive shown powered the fluid drive for the turret traverse mechanism.
▼

Final drive with double spur gear reduction to the drive wheel.
▼

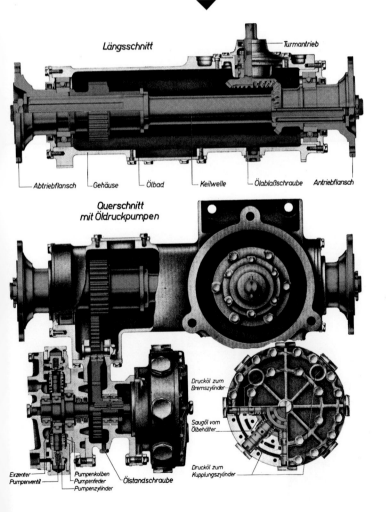

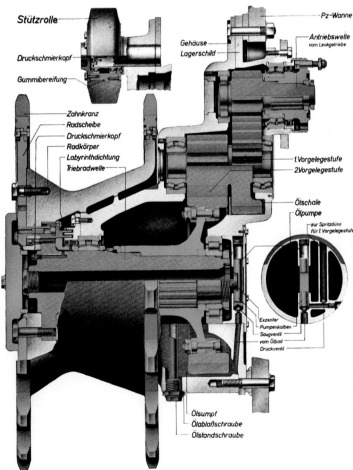

Triple plate dry clutch for transferring the power from the engine to the transmission.

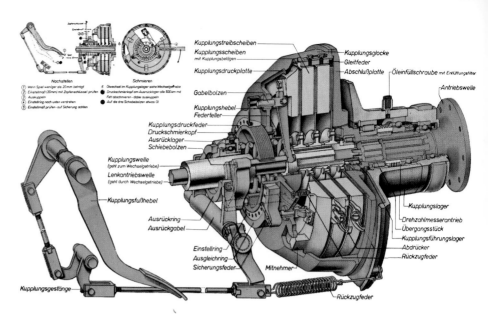

Manual synchronized transmission with seven forward gears and one reverse gear.

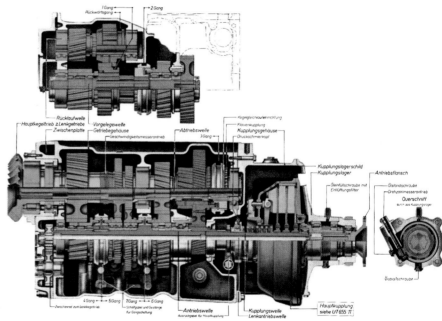

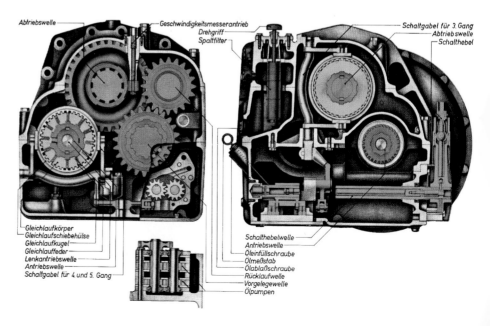

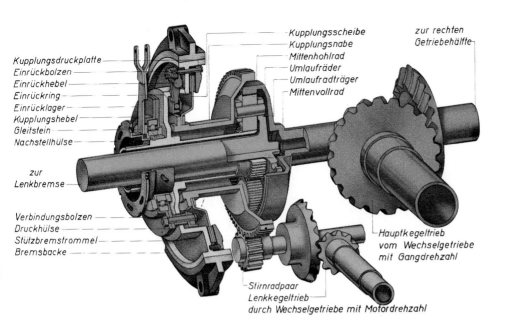

Single-radius controlled differential regenerative steering mechanism: main drive and steering drive as well as brakes for activating the fixed steering radius.

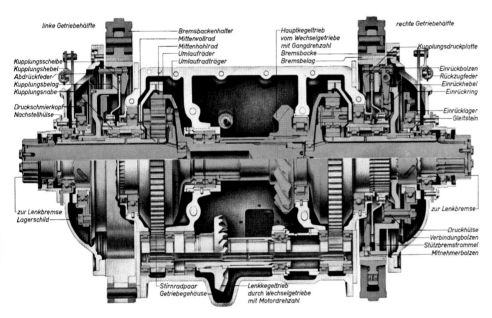

Sectional view of the steering mechanism.

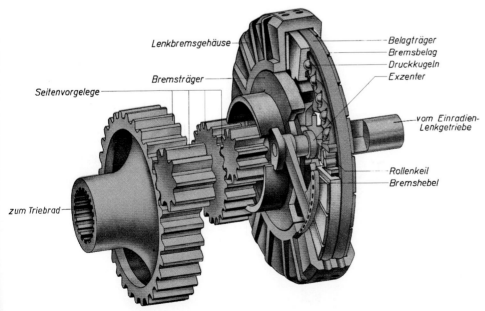

Solid disc brakes, used as steering brakes for tight radii and also as primary brakes. Spur gear reduction for the final drive.

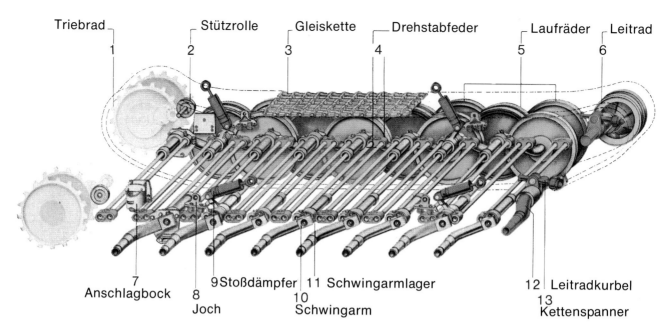

Triebrad	Stützrolle	Gleiskette	Drehstabfeder	Laufräder	Leitrad
1	2	3	4	5	6

7 Anschlagbock
8 Joch
9 Stoßdämpfer
10 Schwingarm
11 Schwingarmlager
12 Leitradkurbel
13 Kettenspanner

Double torsion bar suspension made possible a large vertical stroke (total of 510 mm) for the road wheels. This meant that during cross-country travel the vehicle was able to absorb even hard shocks well.

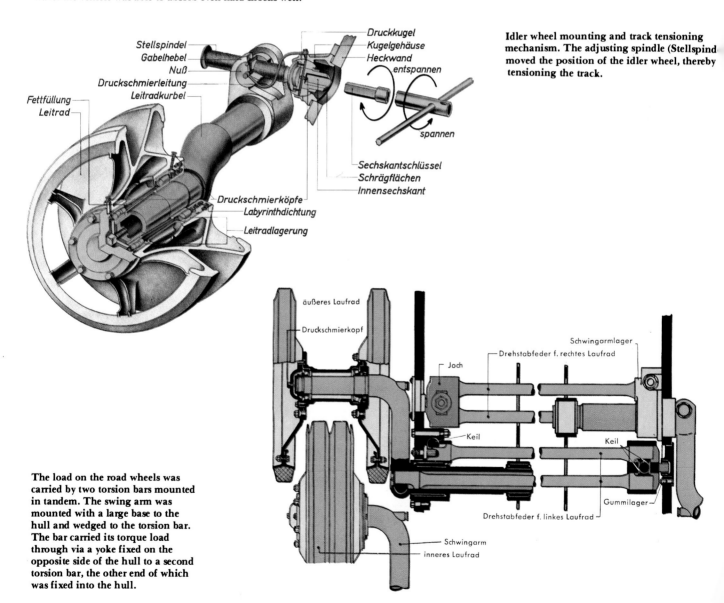

Idler wheel mounting and track tensioning mechanism. The adjusting spindle (Stellspind) moved the position of the idler wheel, thereby tensioning the track.

The load on the road wheels was carried by two torsion bars mounted in tandem. The swing arm was mounted with a large base to the hull and wedged to the torsion bar. The bar carried its torque load through via a yoke fixed on the opposite side of the hull to a second torsion bar, the other end of which was fixed into the hull.

140

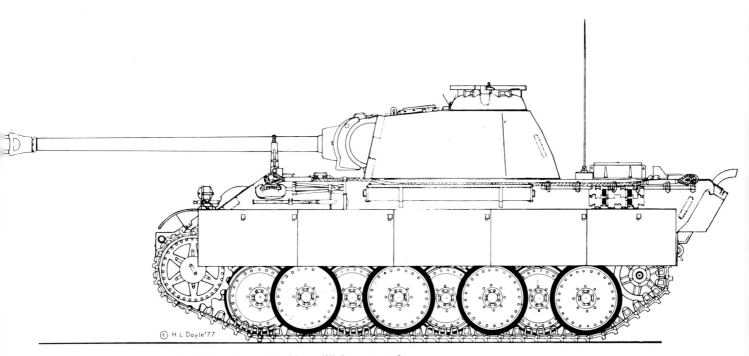

Panzerkampfwagen Panther (Ausf. G) (Sd.Kfz.171) model with modified gun mantelet.

mounted crossways at the rear of the engine compartment on a few vehicles.

On 5 March 1944 Hitler decided to make use of the cave at Leitmeritz on the Elbe (Sudetenland) for the anticipated assembly of tank engines. He demanded that the facility should be arranged so that, depending on the need — if a manufacturing plant should be knocked out of commission — Maybach HL 120 and HL 230 engines could be built there.

At the same time efforts were to be undertaken to press forward with the development of an air-cooled diesel engine in order to be able to manufacture these at the site. The Generalinspekteur der Panzertruppe Generaloberst Guderian and his chief of staff Oberst Thomale held a meeting on 30 March 1944 at MAN, Nuremberg, which primarily concerned the engine and final drive of the Panther. Guderian explained that the situation on the Eastern Front was profoundly influenced by the shortage of tanks and, in particular, criticized the lack of spare parts. Many tanks with relatively light damage were forced to be blown up or worse, were allowed to fall into the hands of the enemy without a struggle. The difficulties with the Maybach engine had been reduced significantly. Guderian and Thomale were surprised to learn that MAN had installed a BMW 132 Dc radial aircraft engine in a Panther for testing. The narrow confines of the Panther's engine compartment were ideal for the installation of a compact radial engine. A large-diameter blade fan mounted in front of the engine, its paddle blades in front of the cooling ribs of the cylinders, provided adequate cooling. The tests were conducted with satisfactory results. They were later broken off when serious thought was given to the accessibility of the lower cylinder heads, and the fact that their spark plugs and vents could only be practically reached by removing the engine. MAN made the suggestion that this testing be given a wider basis and recommended that no less than 20 Panthers be equipped with this engine.

There was still great interest in a diesel engine. With a view to the air-cooled 700 hp MAN-Argus diesel engine it was considered that large numbers of these engines could be made available no earlier than two years hence. The MAN-Argus engine had 16 cylinders and was arranged in an H form with two crankshafts.

The mortising of the armor plating on the welded hull was dispensed with for reasons of simplification, and the first test drive using a non-mortised Panther hull was conducted on 2 April 1944. It was not expected there would be any significant problems. At least 110 Panthers were demanded of MAN for the month of April 1944. In reality, 105 were delivered. In doing so, this offset the March loss due to the conversion to the G series — even though MAN had lost critical days at the beginning of April as a result of this conversion. In order to meet the demanded increase to 130 vehicles per month 22 pieces of stopgap machinery were needed, which MAN did not

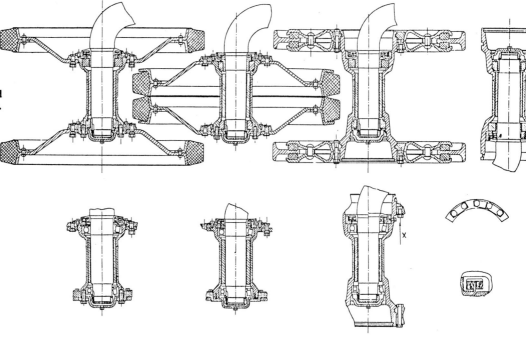

Both the old-style and newer road wheels could be installed interleaved without difficulty. The sketch shows a comparison of the old and new road wheels.

Right and below: A resilient all-steel running gear was installed for the first time on a Panther G, chassis number 121052. As previously, the road wheels had a diameter of 860 mm.

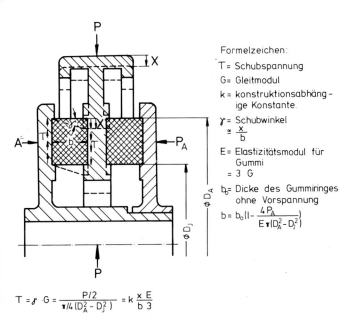

Formelzeichen:
T = Schubspannung
G = Gleitmodul
k = konstruktionsabhängige Konstante
γ = Schubwinkel $\cong \dfrac{x}{b}$
E = Elastizitätsmodul für Gummi = 3 G
b_0 = Dicke des Gummiringes ohne Vorspannung
$b = b_0\left(1 - \dfrac{4 P_A}{E \pi (D_A^2 - D_J^2)}\right)$

$T = \gamma \cdot G = \dfrac{P/2}{\pi/4 (D_A^2 - D_J^2)} = k \dfrac{x}{b} \dfrac{E}{3}$

$P = k \dfrac{E \cdot \pi (D_A^2 - D_J^2)}{6 \, b_0 \left(1 - \dfrac{4 P_A}{E \pi (D_A^2 - D_J^2)}\right)} \cdot x$

Höhenfederung des Laufrades mit zwei Gummiringen

$E = \dfrac{1000}{d_1}$, wobei d_1 für 10 kg/cm² gemäss DIN 53511, Bl. 4, gemessen wird.

Data for the resilient road wheels.

Forschungsinstitut Dr.-Ing. G. Oppel, München, Franz-Josefstr. 41.	
Berechnungsunterlagen für gummisparende Laufrollen.	
Ableitung der Formeln für die Höhenfederung der Rolle mit zwei Gummiringen.	9.1.1943

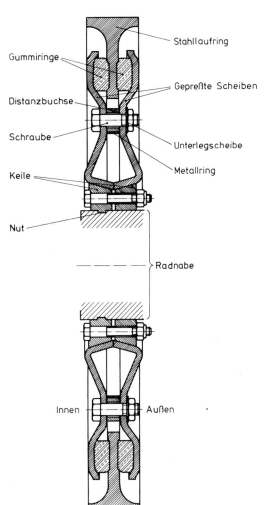

Cutaway of an all-steel road wheel.

BMW 132 Dc aircraft engine for installation in the Panther tank. Fan side with drive from the crankshaft to the gearbox. First model, from 1943. For the Panther the engine delivered 520 hp at 2000 1/min.

Carburetor side of the BMW 132 Dc 9 cylinder radial engine, which equipped two Panthers for factory testing.

have.

On May 13th, 1944 Hitler was finally prepared to have the Panther evaluated for propaganda purposes.

The Zahnradfabrik Friedrichshafen was awarded a contract in the amount of RM 42,000 on 16 September 1944 for the further development of the AK 7-200 transmission. Among other things, this was to include:
— installation of the AK 7-200 in a Tiger with normal and special gear ratios
— installation in the Panther II with Henschel L 801 steering mechanism
— hookup to a double clutch drive of MAN design
— hookup to the Renk hydrostatic steering unit
Testing based on this contract had begun on 5 March 1943.

In October 1944 the Heereswaffenamt also awarded a contract in the amount of RM 8816 for the continued design development of the AK 7-200 gearbox for simplified production and compatible with diesel engines of up to 1000 hp, suitable for either the Panther or the Tiger.

In view of the difficulties involved in the timely manufacture of alloy armor plating, Hitler agreed on 22 June 1944 to substituting the rear plates on the Sturmgeschütze and, if necessary, the tanks as well with alternate quality material. This would be done in order to make the needed material available for the frontal armor. Hitler was cognizant of the fact that both the manufacturing industry and the offices responsible for carrying out this change were aware of the extraordinary burdens inherent in such an undertaking. Hitler expected that the significant developments already being made with regards to alternate quality materials for the purpose of achieving parity with the current armor quality would continue at an energetic pace.

In July of 1944 five Panthers were delivered to Hungary.

On 12 October 1944 Hitler was provided with an overview of the maximum gun ranges for all tanks, based on their optics and gun elevations. Hitler demanded that, if necessary, the tanks were have the capability to be used as artillery pieces, and that the necessary prerequisites for this were to be drawn up in the near future as an emergency outline.

On 28 November 1944 Hitler instructed that the optics on the tanks be equipped with a windshield wiper.

A cost assessment from Daimler-Benz Berlin-Marienfelde dating from August of 1944 gives an indication of the group price for the entire Panther production. A report filed by the OKH/Heereswaffenamt/WuG 6 from 5 July 1944 established the batch price

Some G models had the container for the gun cleaning equipment mounted crosswise over the engine compartment.

for a single Panther chassis and superstructure for a six month period beginning 1 June 1944 as follows:
for the chassis: RM 62,000
for the superstructure: RM 14,000
total: RM 76,000

The following picture resulted from the latest calculations (700th to 1000th vehicle):

	chassis	superstructure	total
Batch cost	62000.–	14000.–	76000.–
./. 0.5% for employees			
Berlin sales	310.–	70.–	380.–
./. tax	1240.–	280.–	1520.–
net income	60450.–	13650.–	74100.–

Recalculation (700th to 1000th vehicle)

	chassis, allotted	actual	superstructure, allotted	actual	total, allotted	actual
materials	33634.-	33634.-	10334.-	10334.-	43968.-	43968.-
wages	3120.-	2839.-	805.-	733.-	3925.-	3572.-
general expenses	14820.-	13059.-	3824.-	3372.-	18644.-	16431.-
total	51574.-	49532.-	14963.-	14439.-	66537.-	63971.-
3% risk	1547.-	1547.-	449.-	449.-	1996.-	1996.-
total internal costs	53121.-	51079.-	15412.-	14888.-	68533.-	65967
profits	7329.-	9371.-	1762.-	1238.-	5567	8133.-
% internal costs	13.8	18.3	11.4	8.3	8.1	12.3

It is interesting to note that the estimated calculation included a 5% extra "security" charge for materials and a 10% surcharge for production wages. The surcharge for materials was necessary in order to offset the constantly changing suppliers and their different prices.

The extra charge for wages provided a certain reserve. The 3% risk surcharge afforded a safety net against the design changes which were constantly taking place. Even though the overall batch price was considered adequate, the cost of the chassis turned out to be quite good while the superstructure resulted in a significant loss. The total cost for a Panther (minus weapons and radio equipment) amounted to RM 117,100.*

On August 10th 1944 MAN reported that there had been no problems encountered in the manufacturing of the turret bearing race up to that point.

The quota program for September 1944 was set at 150 Panthers for MAN. 140 vehicles were actually delivered, in spite of the fact that the assembly line was stopped six times during the month of September due to lack of hulls, engines and torsion bars.

At the end of September 1944 it was discovered during the final acceptance of a number of Panthers that some of the torsion bars Röchling had supplied were too weak. In accordance with a directive from the OKH/Heereswaffenamt/WuG 6 17 Panthers with defective torsion bars were sent to the Grafenwöhr training grounds, where the faulty torsion bars were immediately replaced. 350 hours were required per vehicle.

The Luftwaffe received 75 Panther tanks and 2 Bergepanther recovery vehicles in 1944/1945.

In 1944, Maschinenfabrik Augsburg-Nürnberg demanded compensation in the amount of 3 million Reichsmarks from the Heereswaffenamt/WuG 6 for the development of the Panther, the pre-production work and bringing the Panther up to series standards as well as the further development of the vehicle. The Heereswaffenamt took the viewpoint that such a demand fell entirely outside the framework of compensation normally granted and considered actually reasonable.

The request was denied for the time being, since MAN had incurred no additional developmental costs and since every single hour spent on design as well as the cost of all prototype models were paid for by the Heereswaffenamt down to the last pfennig. MAN had apparently realized such high profits through its own assembly of over 1700 Panthers (stemming from its development program) up to that point that these profits should have been considered satisfactory. Only the quantity of material actually needed to meet normal demands, not the quantity of war material, was permitted to be considered when calculating the license fees. After heated negotiations it was finally agreed that the Heereswaffenamt was to reimburse Maschinenfabrik Augsburg-Nürnberg the sum of 1.2 million Reichsmarks for developmental costs.

War contract SS 4911-0210 5902/43 was assigned to Krupp-Gruson for the production of 1529 combat-ready Panther tanks. In October 1944 this contract — initial deliveries of which were to begin in April 1945 — was canceled. Krupp was informed by the Heereswaffenamt that it would receive partial payment both for costs already incurred as well as for the outstanding balance.

On 24 October 1944 Henschel turned blueprints over to MAN showing the planetary gearing system for the steering unit using sleeve bearings. The study was necessary since neither ball nor roller bearings were now available in sufficient quantities.

A meeting of the Hauptausschuss Panzer on 22 November 1944 chiefly concerned itself with converting the ball and roller bearing elements in armored vehicles to sleeve bearings. Accordingly, one Panther at MAN and

---

* Comparison: Panzer III RM 96163.-, Panzer IV RM 103462.-, Tiger I RM 250800.-

two at Krupp were converted for testing. These testbeds demonstrated acceptable results. Sleeve bearings were made available to the plants for the track adjusting system beginning in January of 1945 and for the running gear beginning in February of 1945. Without waiting for the results of testing from a pre-production series, sleeve bearings were to be introduced into the ongoing Panther production on short notice. Contracts were awarded to Braunschweiger Hüttenwerke for developing babbitt sleeve bearings for 1200 Panthers. Some of the raw materials for these bearings were supplied by MAN to Braunschweig. The steel rings for the 1200 fittings were produced by the individual vehicle manufacturers.

20 of the 45 ball and roller bearings in the AK 7-200 gearbox were planned to be converted to sleeve bearings. However, transmission testing using Glyco bearings had not been concluded by the war's end.

Using a Panther Ausf. G (chassis number 120 303) Henschel conducted comprehensive testing using a Dräger air filtration system. The Schutzlüftungsgerät 0.6-2 served the purpose of supplying the tank's fighting compartment with clean air and enabling the crew to operate inside the vehicle without the necessity of gasmasks. In order for this to occur the tank body must be sealed against gas. When the filtration system was operating, the tank's openings and hatches could not allow any air to enter from the outside. The equipment could also be used for conventional ventilation by bypassing the spatial filter. Its performance capacity was officially rated at 0.6 m³/min, but in actuality this was closer to 0.7 m³/min when operating as a filter, and approximately 1 m³/min when ventilating.

The equipment was developed in such a manner that it could also be installed in non-armored vehicles which had the capability to be sealed. The system was driven off the cardan shaft by a V-belt, which in turn drove the shaft to the air filter. The air which was filtered into the fighting compartment passed through a dust pre-filter (cyclone filter) and two primary filters (coal filters) set crosswise to the direction of travel. An Askania micro-manometer was used to measure the pressure data.

As late as 1944 it was discovered that the design weaknesses in the final drive in the Panzer IV and Panther were constantly leading to high numbers of tank breakdowns, and the needs of the troops — in spite of the greatest efforts by the industry (without affecting series production) — could simply not be met.

It is interesting to examine a compilation of testing contracts for the Panther tank which the Heeresversuchs-

Early on testing was undertaken for protecting the tank crew against poison gas by using an air filtration system. Here is a mockup of a system in a Panther Ausf. D.

Henschel explored the installation of a Dräger air filtration system in a Panther G (chassis number 120303).

Pre-filter box with blower intake for the air filtration system.

Two coal filtration boxes were installed crosswise to the direction of travel.

Drive shaft with belt drive for the ventilation blower.

The ventilation blower was driven by a V-belt, which in turn was powered by a small drive shaft.

stelle für Panzer and Motorisierung, Kummersdorf had received for 1944:*

Month/Year	Contract
6/44	continuous turret drive, Steinrück model
6/44	all-steel resilient road wheels
6/44	Novotext sleeves for swing arm mounts
6/44	Bergepanther armored recovery vehicle:
	1. reinforced final drive (planetary)
	2. cable winch and equipment
6/44	AK 7-200 gearbox
6/44	VMS 125 steering drive with cast steel housing and gear wheels
6/44	steering drive minus ball bearings for sun gear
7/44	P 100 high pressure piston pump (made by Klopp)
	for steering unit
8/44	ball bearings in place of roller bearings for final drive
8/44 steel	Argus brakes with emery coating instead of cast gray
8/44	steering drive (MAN), Argus, Teves and Daimler-Benz manual steering using with steering wheel as well as hydrostatic steering drive
8/44	860/100 DB semperite tires, without steel bands, with imbedded metal sheeting in place of wire cable (drive testing)
9/44	radial ball bearings for idler wheel
10/44	drive sprocket slip rings for towing
11/44	compact turret (Schmalturm)
11/44	sintering iron slide bearings for turret traverse drive and elevation drive

* Graciously provided by Oberst a. D. Dipl.-Ing. W. Esser on 27 September 1976.

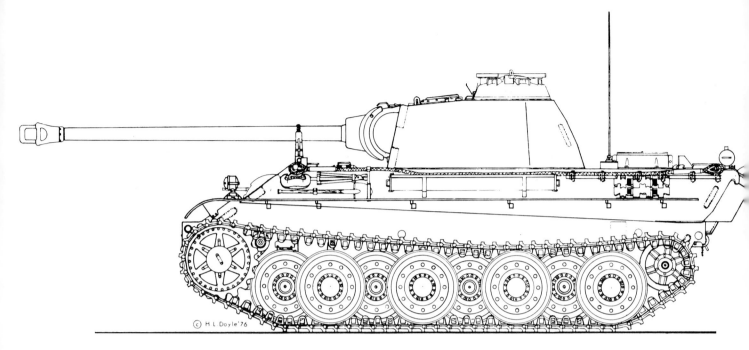

Panzerkampfwagen Panther Ausf. G (Sd.Kfz.171) (with resilient all-steel road wheels)

3768 Panther tanks were manufactured in 1944. On 1 February 1945 another 1216 Panthers were planned for production.

At the end of 1944 Daimler-Benz undertook preparations for the production of the Panzerkampfwagen Panther, Ausf. F (Sd.Kfz.171). Daimler-Benz, in cooperation with Skoda, had developed a new turret for this vehicle, for which the Heereswaffenamt/WaPrüf 6 had laid down the following construction guidelines:

— Eliminating the deflection of shots into the hull roof by changing the design of the turret mantelet.
— Increasing the armor protection without increasing the weight of the turret.
— Reduction of the turret "shot traps" without reducing the interior space of the turret.
— Installation of a stereoscopic rangefinder.
— Replacing the coaxial MG 34 by an MG 42 (7.9 mm caliber, muzzle velocity of 740 m/s, 1500 rounds per

Panzerkampfwagen Panther Ausf. F (Sd.Kfz.171)

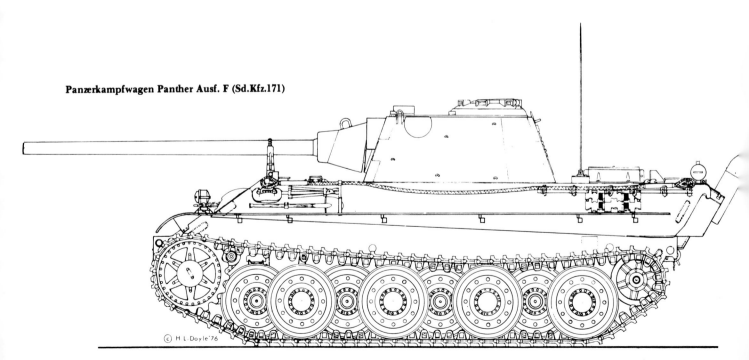

Above: The Panther Schmalturm was planned for the F model of the Panther tank. Here the Schmalturm has been set on a G chassis (chassis number 120413).

Left: The gun was equipped with a muzzle brake which was to be dropped on the production version.

Below: The Schmalturm prototype for the Panther tank.

The Schmalturm shown in these photographs can be seen today at the Royal Armored Corps Center, Bovington, England.

minute)
— Reduction of manufacturing costs for turret.
— Installing the necessary mounts for conversion to a command vehicle and for accepting infra-red equipment. The armament would then consist of: 1 75mm KwK 42 or 44 (L/70); 1 turret mounted MG 42; 1 Sturmgewehr 44 assault rifle mounted in the forward hull; 2 machine-pistol "P" modifications (assault guns with shortened barrels)

It was expected that a 30 to 40% savings in production time over previous models would be possible with the new turret. Despite significantly thicker armor the original turret weight of roughly 8 metric tons was not exceeded. The "shot traps" on the turret were reduced without restricting the interior space inside the turret. The turret traverse ring diameter was not altered. The 75mm KwK was improved by moving the recoil mechanism beneath the gun barrel. As found on the Tiger II a "Saukopf", or pig's head mantelet was utilized. The new gun was designated 75mm KwK 44/1. The gun cradle was no longer a welded construction and the earlier barrel blowout mechanism was now replaced with an additional cylinder installed in the barrel recoil mechanism. This cylinder was operated by the first 420 mm of the barrel recoil.

There were no plans to make use of a muzzle brake even though the initial KwK 44/1 guns were equipped with one. The greater recoil force of 18 instead of the previous 12 tons had been taken into account.

The turret frontispiece was made of rolled steel plate which, compared with the older Panther gun mantelet, required no specialized manufacturing process.

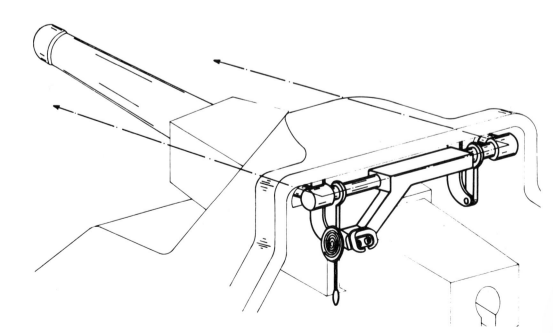

Sketch showing installation of the rangefinder in the tank (diagrammatic sketch).

Skoda, in conjunction with Krupp, carried out the necessary changes to the gun for the Schmalturm.

* According to information graciously supplied by Dr.-Ing Gaertner.

During combat, the Panzertruppe required the capability of engaging enemy tanks at a distance of 3000 meters. Although exact ranging was not required at shorter distances due to the high muzzle velocities of tank guns, there was certainly a pressing need for range-finding in the region beyond 2000 meters in order to determine the angle of elevation more accurately. It was also shown from experience in North Africa that estimating ranges was particularly difficult in desert terrain due to a general absence of reference points.

Therefore, in 1942 the development of rangefinders for tanks was begun at the Zeiss company, Jena, at the instigation of the Heereswaffenamt.

Since the dimensions of individual tank types often varied it became necessary to fit the base length of such rangefinders and their mounts to the existing dimensions of the turret. It was agreed that the rangefinder would be given to the gunner instead of the commander.

A combination of rangefinder/gunsight was proposed. A magnification of roughly 15x was necessary for the rangefinder due to reasons of accuracy, which led to a relatively small field-of-view of approximately 4 degrees. But since a large field-of-view of 28 degrees was needed for the gunsight, the requirement for a combination device was dropped and the method was deemed satisfactory whereby the measured range was automatically transferred by mechanical means to the target marker in the gunsight. In addition, the eyepiece for the rangefinder was placed directly next to the ocular of the gunsight so as to make it easier for the gunner.

The mounting of the rangefinder in the tank was considered very carefully, since peak acceleration in the magnitude of 10 to 15 g occurred during travel across medium rough terrain. When suffering hits, this acceleration was multiplied, even when these hits did not put the tank out of commission. By mounting the rangefinder in two swing supports it was possible to absorb the shock caused by hits to the point where the acceleration was reduced from 1000 g to 60 g. Practical testing reinforced that illuminated marker stereoscopic rangefinders held fast under such shocks without becoming knocked out of alignment. The weight of the rangefinder was offset by a spring to the point where balance was maintained at any position of the support arms.

The following rangefinders were developed for tanks:
a) for Panther with Schmalturm
rangefinder: b = 1.32 m; m = 15x, 4 degree field-of-view, aperture 30 mm. With this turret consideration was given to using a rangefinder during the development phase. For the two objectives of the rangefinder two slits were made in the turret, which could be closed off when not in use.
b) for Tiger II
rangefinder: b = 1.6 m; m = make 15x, 4 degree field-of-view, aperture 30 mm. Since the turret was not originally planned for the mounting of a rangefinder, two side "ears" were affixed to the side of the turret protecting the ends of the rangefinder which protruded from the turret. In addition, it was necessary to provide for two objectives slits which allowed the rangefinder an elevation range of -8 to +15 degrees.

Finally, a clarification in the form of an error table for optical rangefinders is provided. It should be noted that the error-of-measurement E of a stereoscopic rangefinder is proportional to E E^2. The error-of-measurement therefore grows with the square of the distance E to be measured. On the other hand, measurement accuracy is improved by extending the base b as well as an increase in the magnification m

The following items were installed in the turret: a close-combat defense weapon to the right rear of the turret roof, a newly-designed commanders cupola and an escape hatch in the rear turret wall. Based on experience gained in Russia, a pistol port was envisioned for the rear turret wall. A rangefinder was mounted immediately behind the front plate of the turret. This was a stereoscopic design of the firm of Zeiss having a base of 1.32 meters and fixed target markers, approx. 15x magnification and was to be used as a basic rangefinder and not a targeting device. It was tested in a few vehicles where it was found that its main drawback was in the shockproof mounting and protection against adjusting slippage when firing. In the end, the matter was sufficiently dealt with by using a spring mount; however, the rangefinder never went into series production*. The periscope for the loader was dropped.

	Magnification devices for infantry and artillery, including devices for tank commanders b = 0.9m; m = 14x 3 degrees E		Panther with Schmalturm b = 1.32m; m = 15x 4 degrees E		Tiger II b = 1.6m; m = 15x 4 degrees E	
	theoretical	actual (theor x3)	theoretical	actual (theor x3)	theoretical	actual (theor x3)
m	m	m	m	m	m	m
1000	4	12	2.5	7.5	2	6
2000	16	48	10	30	8	25
3000	36	108	23	68	19	56
4000	63	189	40	120	33	99
5000	99	295	63	189	52	155
6000	142	425	91	273	74	222
7000	194	580	124	372	101	303
8000	253	760	160	480	132	396
10000	400	1200	250	750	200	600

The inner ring of the turret bearing race was now an integral part of the turret ring, whereas previously it had been a separate piece. The hydraulic turret traverse drive was generally the same as on earlier Panther models. When the vehicle was situated at steep angles the loader could assist the gunner in rotating the turret by using a handwheel.

By eliminating the two counterbalance gears and multi-plate load clutch on the earlier turrets it was possible to produce a cheaper, smaller and lighter turret traverse. The commander could independently traverse the turret onto a target by using a separate connection to the hydraulic traverse drive.

The elevating drive was arranged under the gun and connected to the framework of the turret basket on one side and with the barrel cradle on the other. The new manual elevating gear was lighter, more compact, and produced at a cheaper rate.

The new commander's cupola presented less of a target thanks to a smaller silhouette, was of cast steel design and had seven openings for shell-proof and waterproof periscopes. A ring for either a scissors-type scope, an AA machine gunmount or a mount for infra-red night fighting equipment was fixed to the cupola. A turret azimuth indicator was installed on the lower part of the cupola.

The installation of a coaxial turret-mounted MG 42 required a completely new mount. It was attached to the gun cradle and consisted of a barrel support with return spring, a rear receptacle for the machine gun, a forward barrel clamp and an adjusting mechanism. Two ammunition containers were fixed beneath the machine gun, one for the ammunition belts and the other for the spent cartridges. Conversion to the MG 42 was necessary because the MG 34 would no longer be produced.

The receptacle for the new TZF 13 turret gunsight was to the right of the barrel cradle. There was now a slit in place of the circular vision opening in the turret front plate. The rear turret hatch was nearly the same as that on earlier models; however it was no longer produced separately, but from the part cut out of the rear plate — thereby conserving armor.

The turret floor was fixed to the inner turret ring and had tubular framing for securing the floor plate, the gun elevating mechanism, the hydraulic traverse motor and a container for empty cartridges. The right side of the floor plate was hinged to permit accessibility to additional ammunition and the batteries.

The turret ventilator was no longer in the turret roof; instead it was fixed near the turret race ring on the right side of the turret. The ventilator provided for the removal of powder gasses and for ventilating the fighting compartment. Metal shafts were used in place of the previous metal tubing.

There were plans to later outfit the turret with the 88mm KwK 43 L/71 in conjunction with a stabilized sight. These plans remained in their initial stages; Daimler-Benz had only been able to construct a wooden mockup of the turret.

This is a compilation of the most significant changes for the Panther-Schmalturm:

Main armament: 75mm KwK 44/1, developed by Skoda, Pilsen, muzzle kinetic energy to 285 mt, weight of gun 1920 kg

Turret gunsight: TZF 13, articulated, monocular magnification 2.5x and 6x, 28 degree field-of-view at lower magnification 12 degrees at higher magnification. Developed by Leitz, Wetzlar.

Rangefinder: Horizontal, stereoscopic, 1320 mm base, 15x magnification, angle-of-view 4 degrees. Developed by Zeiss, Jena.

Turret traverse drive: Traverse speed when manually operated: 360 degrees in 4 minutes, when hydraulically operated: 360 degrees in 30 seconds. Developed by Daimler-Benz, Berlin-Marienfelde.

Hydraulic turret: 6 hp at 800-4200 1/min, traverse mechanism developed by Böhringer GmbH.

Gun elevating mechanism: -8 to +20 degrees. A single turn of the handwheel gave 0.4 degrees elevation.

Other than the reduced turret size, the following additional improvements were planned for the F model of the Panther tank:
— hull top thickness increased from 12 to 25 mm.
— new design for the driver's and radio operator's hatches. The hatch covers were easily raised and swung out to the sides.
— The radioman's machine gun was to be replaced by a ball mount for an MP 44.
— The mounts for the radio equipment were designed in such a manner that conversion from a tank to a command vehicle could by undertaken by the troops in the field. Only the shortwave transmitter was installed in the chassis; remaining equipment was to be found in the turret. Each vehicle was given two antenna sockets.
— Each vehicle was given mounts for the FG 1250 infrared night vision equipment, so that the IR night sight and searchlight could be affixed by troops in the field. The installation of this additional equipment in no way affected the combat capability of the vehicle during daytime operations.
— In place of the earlier interleaved running gear plans called for staggered all-steel running gear.

As on the G model, the equipment for submersion was dispensed with; a heating system for the fighting compartment was planned.

Further development of the gun resulted in the 75mm KwK 44/2 with loader, which was also carried out by Skoda. The barrel length remained unchanged at L/70, with a recoil of 430 mm. Three of these test guns were built and sent to the Rheinmetall-Borsig test range at Unterlüss. It was a mechanical loading apparatus, which was operated by the recoil of the gun. The loader consisted of a framework, fixed to the right side of the gun carriage and holding 4 rounds. By means of a mechanical process, activated by the gun's recoil, a round was lifted up and set into a holder at the end of a pivoting arm. The gun recoil was not affected by this process. After the cartridge had been ejected the breechblock remained open, the holder swung downwards and a tension spring guide forced the round into the breech. The breechblock closed automatically. Closing the breechblock released the pressure on the pivot arm spring, which in turnbrought the holder back up ready to accept another round. At the same time the barrel recoil set the spring guide back in its original position.

This was not a fully automatic loader, but a mechanical rapid-reload device which made possible a rate of fire of up to 40 rounds per minute. In its installed state the gun together with the loader weighed 3400 kg. Since the design of the loader took up considerable space, space inside the turret was very restricted. Preparations were made for installing it in the Panther, Ausf. F.

The Panther Ausf. F was no longer to be built at the Maschinenfabrik Augsburg-Nürnberg. Instead, Daimler-Benz delivered an F model chassis in 1945, with interleaved all-steel road wheels and the turret of an Ausf. G. The Schmalturm was set on the chassis of an Ausf. G and tested at the Verskraft Kummersdorf. At a meeting of the Heereswaffenamt on 25 January 1945 a desire was expressed to build the ZF AK 7-200 gearbox as a five or six gear transmission as soon as the new 850 hp fuel-injected Maybach HL 234 internal combustion engine or the two-stroke Deutz diesel engine (T8M118) became available.

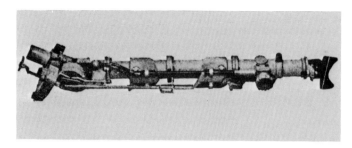

TZF 12a (turret gunsight) for the tank. Monocular with magnification adjustments. It was the last sight put into production.

TZF 13 (turret gunsight). Binocular with magnification adjustments and headpiece.

MAN had changed the new epicyclic final drive to the point where the maximum speed was now 45 instead of the earlier 55 km/h. Increasing the engine r.p.m. had resulted in a 45% increase in unnecessary torque in first gear with the current gearboxes. MAN therefore recommended dropping the first gear entirely and slightly increasing the ratio of second gear. The previous second gear would then be for creep speed; third gear would be for low gear. Since ZF was already in possession of a contract for simplifying the AK-200 gearbox, and since the production of two test gearboxes (type AK 6-200) was already underway, the additional omission of first gear would result in a welcome further simplification. WaPrüf 6 took the view that, with the installation of the 850 hp Maybach HL 234 engine, there would be no problem with a 5 gear transmission. However, the installation of the Deutz diesel engine*, with a torque of 190 mkg and a total reduction ratio of 9, would require the installation of an intermediate Z 200 gear drive for the AK 7-200 gearbox. ZF was given a contract for five of these.

Beginning with the TZF 4, the development of turret gunsights (TZF) lay in the hands of the firm of Leitz in Wetzlar. Previous types — especially the TZF 2 — were developed by the Zeiss firm in Jena. The Leitz models were superior due to the use of total reflective prisms. In the mid-1930s the surface reflectors used by Zeiss were manufactured with a relatively poor reflective quality, so that there was a 50% loss of light caused by the mirror alone.

On the other hand, the arrangement of the scale markings was more in line with what the troops wanted. Using a ranging scale with up to four types of ammunition prevented errors, where only one or two ranges were easily legible on the sighting mark. Both designs were so-called "kinked", or articulated sights, the operative part of the sight being fixed to the gun. All optical parts behind the image plane were for sighting, were therefore sensitive to adjustment and could be spatially mounted.

The turret gunsights in German tanks were initially of a monocular design; later binocular sights were developed at the direction of the Heereswaffenamt WaPrüf 8. The increased expense in no way correlated with the results and these were later given up. In addition, turret sights were developed with adjustable magnification, such as those used in the Tiger and Panther tanks. A drawback was noted in that with increased magnification the field of view became smaller. It was sufficient to switch to a larger magnification for a short period of time in order to better

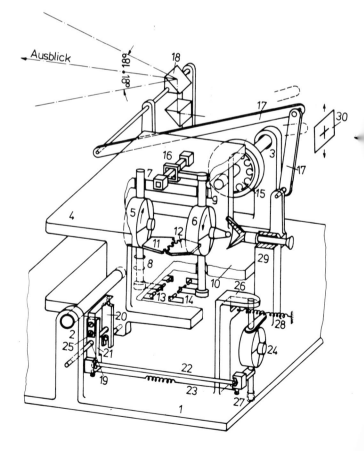

---

* Deutz T8 M118, water-cooled 2 stroke diesel 8-cylinder V-form — 170x180 bore/stroke — 32.3 liter swept volume — 700 hp at 2000 1/min.

acquire the target, but for normal operations the more important large field-of-view was retained.

The stabilized gunsight was based on foreign examples. Aside from the Russians, the Americans had also made use of this type of equipment. In principle, it revolved around a gyro controlled prism in front of the lens sight.

The German copy required a change to the lens socket for the sight. It was assigned to the firm of Kreiselgeräte Berlin (branch of the Askania-Werke, Berlin) and Leitz, Wetzlar. A significant improvement was the use of a double gyro to offset the drift of the gyro axis (precision). The design was completed in a short period of time. As opposed to the Russian design, there was no attempt to make it compatible for firing while on the move; the technology of the day could not guarantee a constant lag time. A prototype of the new sights was made available in a relatively short period of time, and a tank equipped with the sight was demonstrated in the autumn of 1944. It was the final model of sight developed by Leitz prior to the arrival of American occupation forces in Wetzlar on 28 March 1945. Approximately 800 of these pieces were transported into the Ruhr pocket just prior to the appearance of the Americans.

Since the fall of 1944, ever-increasing critical shortages of raw materials and damage to the manufacturing plants caused by air raids led to drastic production difficulties. Due to the dropoff in Panther production in the last months of 1944, a considerable amount of material had piled up at MAN because of the shut-down of some of the suppliers (e.g. hull, turret housing) — enough for the manufacture of roughly 400 vehicles. Because of the loss of certain critical parts the production of Panther and Tiger tanks was to cease at the end of May 1945. This was the result of an inspection on 19 February 1945 within the framework of an emergency program established in January of 1945.

The tank companies were paid only 90% of the vehicle price by the Heereswaffenamt upon delivery of the finished vehicle. In addition, deliveries of the tanks had fallen off considerably. A large percentage of the suppliers, however, continued to produce according to contract and required payment. This problem could only be partially alleviated by courier transmission of funds from the Heereswaffenamt. The MAN production plan anticipated 70 Panther vehicles for February 1945, March — 100, April — 120 and May —126.

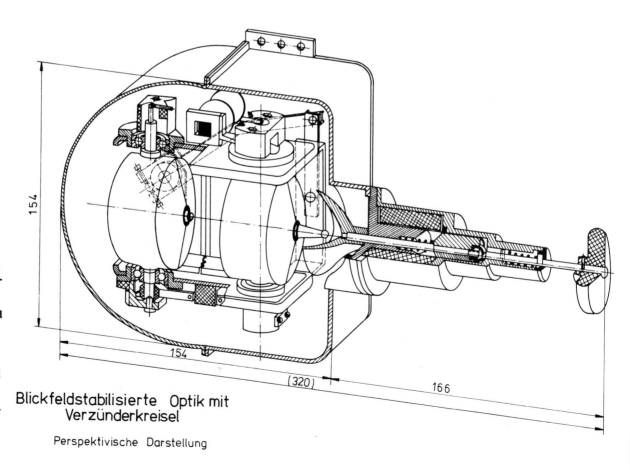

Field-of-view stabilization for the gunsight with gyro-stabilizer. In principle, two prisms were utilized in front of the telescopic lens — the lower prism fixed into the telescopic shaft and the prism controlled by the twin gyro-stabilizer installed above the shaft.

Blickfeldstabilisierte Optik mit Verzünderkreisel

Perspektivische Darstellung

It was not possible to even approach these numbers. In reality only 441 Panthers were produced by all companies involved in Panther manufacturing from January 1945 to the beginning of May 1945.

On 3 March 1945 the majority of the Heereswaffenamt/WaPrüf 6 members were convinced by MAN regarding the status of production and the developmental work on the Panther in view of the emergency program. Research was done to determine the feasibility of continuing Panther production, since this tank offered technically refined advantages and was one of the best available combat vehicles. The emergency program carried the same provisions as before, including retaining the current labor force, and included further development and research on the Panther such as:

hydrostatic regenerative controlled steering,
electrically driven transmission,
fuel-injected engine, air-cooled diesel engines,
recoilless guns for the Jagdpanther,
stabilized gunsight,
infra-red night fighting device.

There were plans for using a running gear with parts common to both the Tiger and Panther tanks*. The running gear of these vehicles had demonstrated superior technical qualities. However, when damaged by mines the double torsion bar suspension of the Panther often required the aid of a welding torch to be removed.

This, along with the requirement for crew escape hatches in the hull floor and the necessity of replacement parts for two different vehicles, led to the exploration of different types of suspension.

At the beginning of 1944 Professor Dr.-Ing Lehr expressed his desire to develop a suspension based on conical disc springs. He felt that all resources should be devoted to the development of this system.

In converting from torsion bars to other spring types, obviously only the suspension having best qualities were considered. Conical disc springs were found to be adequate. The sprung weight of the Tiger was set at 60 tons, while that of the Panther was 40.8 tons. The track length in contact with the ground was 4095 mm for the Tiger and 3850 mm for the Panther. A study group under the direction of Oberingeneur Karl Schindler explored the possibility of supporting the Panther on six road wheels and the Tiger on eight wheels per side, with both having nearly the same load on the bogie wheels. The main springs were cushioned by supplementary springs concentrically arranged within the main spring, based on the stationary load on the wheels — 3400 kg for the Panther and adjusted for the greater road wheel load of the Tiger at 3750 kg.

The load-bearing capacity of the spring column was set by the length of the spring housing; the greatest possible travel stroke (measured at the road wheels) was 170 mm upward and 130 mm downward for a total of 300 mm.

Hydraulic absorbers inside the main cup spring columns were to provide the necessary damping. The spring housing, holding two bogie wheels of the running gear, was arranged symmetrically; the outer and inner road wheel were held in place by two differently shaped wheel hubs.

Running gear data for the Panzerkampfwagen E 50/75; the running gear was planned as an integral unit for the Panther.

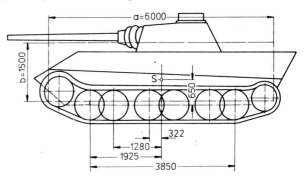

gegeben: $G_f = 40800$ kg; $P_{st} = 3400$ kg; $H_s = 650$ mm; $x = 1{,}15$; $a = 6000$ mm; $b = 1500$ mm; $L_f = 3850$ mm;

Massenträgheitsmoment Θ:

$m = 40800 : 981 = 41{,}65$ kg s²/cm;

$i = x\sqrt{\dfrac{a^2+b^2}{12}} = 1{,}15\sqrt{\dfrac{600^2+150^2}{12}} = 206$ cm

$\Theta_S = mi^2 = 41{,}65 \cdot 206^2 = 17{,}6 \cdot 10^5$ cm kg s²

$\Theta = \Theta_S + mH^2 = 17{,}6 \cdot 10^5 + 41{,}65 \cdot 65^2 = 19{,}3 \cdot 10^5$ cm kg s²

Federkonstanten der Hub- und Nickschwingungen:

$C = 113$ kg/cm; $C = 2C_R = 2 \cdot 115 = 226$ kg/cm; $C_H = 6 \cdot C_R = 6 \cdot 226 = 1356 \dfrac{kg}{cm}$

$C'_N = \Sigma (CL^2) = 2 \cdot 226 (192{,}5^2 + 128^2 + 32{,}2^2) = 246 \cdot 10^5$ cm kg

$C_L = m \cdot g \cdot H = 41{,}65 \cdot 981 \cdot 65 = 26{,}5 \cdot 10^5$ cm kg

$C_N = C'_N - C_L = (246 - 26{,}5) 10^5 = 219{,}5 \cdot 10^5$ cm kg

Eigenschwingungszahlen:

$n_H = \dfrac{30}{\pi}\sqrt{\dfrac{C_H}{m}} = \dfrac{30}{\pi}\sqrt{\dfrac{1356}{41{,}65}} = 54{,}5$ Schwingungen/min

$n_N = \dfrac{30}{\pi}\sqrt{\dfrac{C_N}{\Theta}} = \dfrac{30}{\pi}\sqrt{\dfrac{219{,}5 \cdot 10^5}{19{,}3 \cdot 10^5}} = 32{,}2$ Nickschwingungen/min

Dämpfung:

$D_n = 0{,}35$; $R = 245$; $r = 86{,}4$;

$D = \dfrac{\varrho_n}{2 \cdot \Theta \cdot \omega}$; $\varrho_n = \dfrac{D_n \cdot \Theta \cdot \pi \cdot n_N}{15} = \dfrac{0{,}35 \cdot 19{,}3 \cdot 10^5 \cdot \pi \cdot 32{,}2}{15} = 4{,}6 \cdot 10^5$ kg s cm

---

* Panzerkampfwagen E 50/75, see volume 8 in the German language series "Militärfahrzeuge."

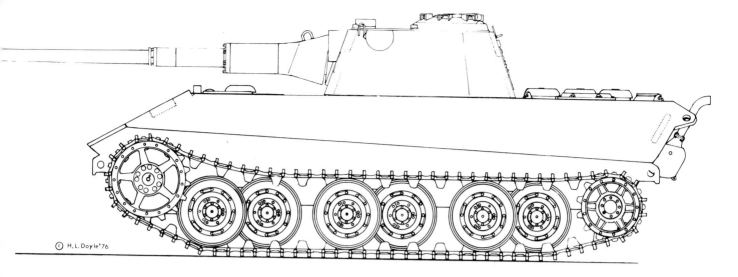

Panzerkampfwagen E 50 (proposal)

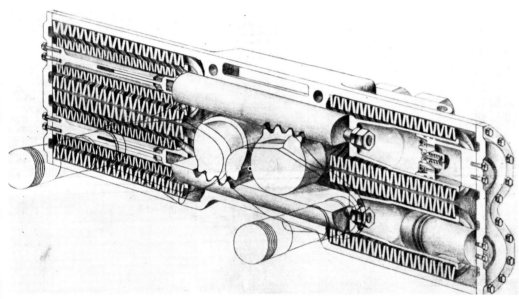

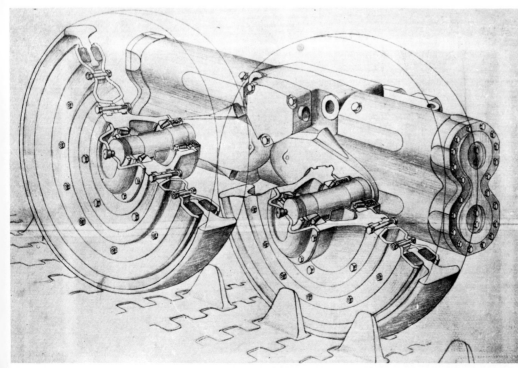

Conical disk spring arrangement for the E 50, a design of the Adlerwerke Frankfurt/Main from 1945. The sketches show the road wheel carriers with spring arrangement in cutaway (above) and standard views.

Drawing of the conical disk spring running gear for the Panzerkampfwagen E 50/75.

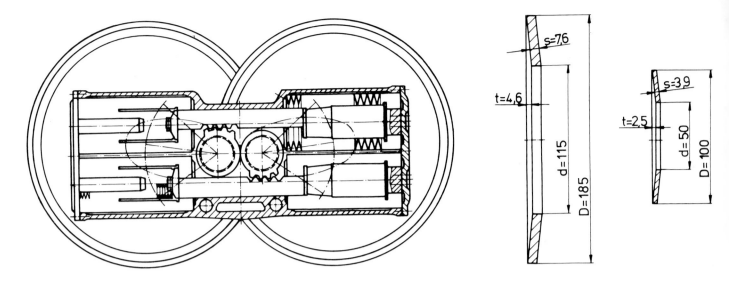

The two sizes of conical disk springs utilized with this type of suspension.

In addition, by eliminating the torsion bars the Panther would now have the lowest possible height.

MAN's total contract from the Heereswaffenamt/WuG 6 for the Panther amounted to 4000 vehicles up to 30 March 1945. This contract, number 210-5901/42, included 12 Bergepanther recovery vehicles. A total of 2038 Panthers were accepted from the Maschinenfabrik Augsburg-Nürnberg itself.

Hitler expected a demonstration by mid-April 1945 of the Panther tank with the 88mm KwK 44 L/71.

An unusual version of the Panther was put into operation during the Ardennes offensive towards the end of 1944. By using metal plates, some vehicles were disguised to look like an American tank destroyer (Carriage, Motor, 90 mm Gun, M 36) and, together with captured American armored vehicles were assigned to special German units. With Allied markings these were expected to cause confusion among enemy forces during the initial phases of combat.

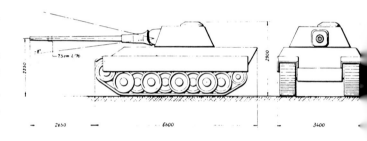

Towards the end of 1944 Krupp experimented with the concept of converting the guns in all German combat vehicles still in use by the troops. The sketches show the Panther proposal (above) with an 88mm KwK 43 L/71 and a Panzerjäger Panther (below) with a 128mm Pak 80 L/55.

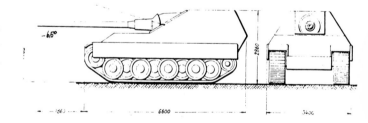

For the Ardennes offensive at the end of 1944 select Panthers were disguised as American tanks for the purpose of causing confusion among the enemy forces.

Gas-powered crankshaft starter attached to a Panther tank.

Heat lamp for the coolant heating element in operating position.

Coolant heating element for the engine.

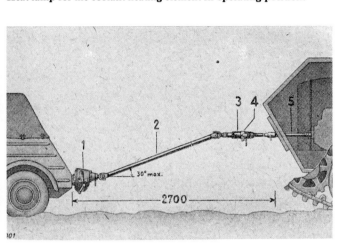

Diagram of the complete starter drive system.

Fixing the guide bridge frame to the tank. 1 cover for starter shaft, 2 retaining bolt, 3 slide lock, 4 guide bridge frame, 5 guide head, 6 retaining bolt.

Mounting of the starter drive on a Volkswagen. 1 hand grip, 2 guide mount, 3 clutch for starting handle, 4 coupling shackle with handle, 5 starting lever with handle, 6 guide bolts, 7 drive housing.

Connection of the coupling sleeve to the tank. 1 auxiliary starter shaft, 2 lock nut, 3 clamping nut. 4 handle, 5 guide markings, 6 universal joint shaft, 7 handle, 8 coupling sleeve, 9 guide bridge frame, 10 centering bolt.

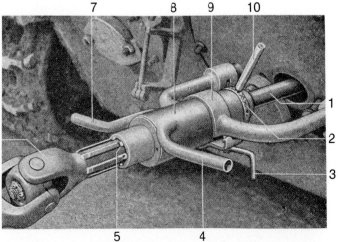

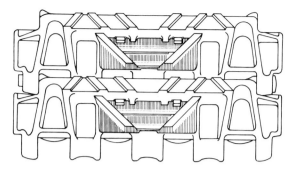

Schneegreifer in Gleiskettenglied eingebaut

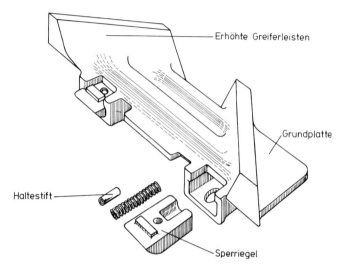

Snow grousers for the Panther tracks.

The gun conversion for all German armored vehicles in front-line service researched by Krupp shows plan number Hln-E 142 dated 17 November 1944 illustrating the mounting of the 88mm KwK 43 L/71 (without muzzle brake) in the original Panther turret.

In 1944 research was conducted using high voltage cables on tanks for protection against close-combat. Two cables ringing each vehicle were planned.

Specialized winter equipment was to ensure combat readiness for the tanks even at an outside temperature of -40 degrees Celsius. The equipment was composed of: a heatable battery with insulated box, a Panzer-Kühlwasserheizgerät 42 (armor coolant heating element) with blowpipe lamp, a fuel-injected starter fluid device, a connector for a gas-powered crankshaft starter as well as non-skid grips (center grips) for the tracks. This equipment restricted speeds to no more than 15 km/h to prevent damage to running gear and road.

The use of a coolant pre-warmer, based on the recommendation of Kriegsverwaltungrat Fuchs (Fuchs Gerät), dispensed with the need of the originally planned coolant feed system.

The gas-powered crankshaft starter basically comprised a plug-in connector attached to the crankshaft via an opening in the vehicle's rear plate. In 1944/1945 Porsche developed a starter drive (Porsche Type 198), which was used in conjunction with a leichte Personenkraftwagen K1, Typ 82 (Volkswagen). When used, the tank to be started and the VW stood end to end and were connected by the starter shaft (which also had slip clutches). Two different levels of speed could be accessed.

For heating the fighting compartment the heated cooling air from the left forward radiator was fed into the fighting compartment. In doing so, the fan wheel of the left radiator group was driven in the reverse direction. A snow remover was available for the Panther when traversing deep snow, which took the form of a snow plow and was fitted to the front of the vehicle.

A total of 6042 Panther tanks were built by the end of the war. Once the majority of the technical difficulties had been overcome, the Panther was, until the very end, the dream vehicle for the German Panzertruppe (a dream which, unfortunately, was often not always attainable).

Captured Panthers were used after 1945 primarily by the French forces. In 1947 a battalion of the 503rd Armor Regiment in Mourmelon was outfitted with 50 Panthers, while additional Panthers (including Jagdpanthers) were also stationed in Satory and Bourges.

French experience with the Panther were put to paper in the report "Le Panther 1947", published by the Ministre de la Guerre, Section Technique de L'Armée, Groupement Auto-Char, which was graciously made available to us. The following are some excerpts from the report:

— The turret traverse drive is not strong enough to either turn the turret or hold it in place when the Panther is on an incline of more than 20 degrees. The Panther is therefore not capable of firing when driving cross-country.
— Elevating the gun is normally simple, but made difficult if the stabilizer — operated by compressed nitrogen — has lost pressure.
— The commander's cupola with its 7 periscopes provides a nearly perfect all-round visibility. Periscopes damaged by shells can be replaced very quickly.
— A scissors periscope with large magnification power was affixed to a bracket in the commander's cupola.
— Aside from his periscope gunsight (which is excellent), the gunner has no other type of observation device. He is therefore practically blind — one of the greatest shortcomings of the Panther.
— The gunsight with two magnification stages is remarkably clear and has its field of view clear in the center. The gunsight enables observation of a target and shells out to over 3000 meters.
— No type of hollow charge ammunition is planned for the Panther.
— The HE shell can be fired with a delay of 0.15 seconds.
— The PzGr 40 had better penetration out to 1500 meters than the PzGr 39, but then its trajectory drops off considerably.
— During rapid rate of fire it is not uncommon to be forced to break off firing when the recoil of the gun has reached its permissible limit (cease fire).

Panther tank in service with the French armed forces (FFI) after the end of the Second World War.

— A rate of fire of 20 rounds per minute is only permitted in exceptional cases when circumstances so dictate.
— When firing off a round the chassis demonstrates no unfavorable reaction, regardless of what position the turret is in.
— Once the commander has located a target, it takes between 20 and 30 seconds until the gunner can open fire. This data, which is significantly greater than that of the Sherman, stems from the absence of a periscope for the gunner.
— The fatigue life of the mechanical parts was designed for 5000 km. The wear on many parts is greater than expected. Track and running gear have a life of 2000 to 3000 km. Tracks break very rarely, even on rocky terrain. The bogie wheels, however, can become deformed when driven hard.
— The parts of the power train (with the exception of the final drive) meet the planned fatigue life. The replacement of a transmission requires less than a day.
— On the other hand, the engine was not operable over 1500 km. The average engine life amounted to 1000 km. Engine replacement accomplished in 8 hours by an Unteroffizier (mechanic by occupation) and 8 men with the aid of a tripod beam crane or a Bergepanther. Main gun can be replaced using the same equipment within a few hours. The German maintenance units performed their work remarkably well
— As a result, the Panther is in no way a strategic tank. The Germans did not hesitate to economically increase the engine life by loading the tank onto railcars — even for very short distances (25 km).
— The truly weak spot of the Panther is its final drive, which is of too weak a design and has an average fatigue life of only 150 km.
— Half of the abandoned Panthers found in Normandy in 1944 showed evidence of breaks in the final drive.
— In order to prevent these breaks it is recommended that the following points be closely observed: when driving downhill and in reverse as well as on uneven terrain to be particularly careful when shifting to a lower gear. In addition, a Panther should never be towed without uncoupling the final drive previously. Finally, under no circumstances should both steering levers be operated simultaneously — regardless of the situation.
— A hollow charge round — regardless of what type — will penetrate armor plating equivalent to its own caliber. It is therefore necessary to use a 105 mm round or, at the very least, an 88 mm round to penetrate the glacis plate of the Panther (Münsingen, 1946)
— A smoke grenade thrown onto the rear deck or the vent openings of the engine will start a fire.
— The running gear is sensitive to HE shells. Calibers 105 mm and greater can render the vehicle immobile (Rammersmatt, 8 December 1944).
— Fragmentation shells or 75 mm rounds which strike in the same spot on the front plate can penetrate it or cause the weld seams to break (Münsingen, 1946).
— No place of the Panther is so armored that it can withstand a "Panzerfaust" or "Panzerschreck."
— In all cases, the great range of the gun should be exploited to the fullest. Fire can commence at a range of 2000 meters with considerable accuracy. The majority of hits were accomplished at a range of 1400 to 2000 meters. The ammunition expenditure was relatively low; on the average the fourth or fifth shot found its mark, even when using HE shells.

Without a doubt, the Panther was a fully combat-capable tank in 1943, which for its day exhibited remarkable performance in regard to its armament and armor.

Yet even German documents showed that it had considerable weaknesses:
— Inadequate for strategic mobility due to the short fatigue life of its engine, which lay between six and seven times the vehicle's range. The Panther cannot cover large distances and must restrict itself to short distances.
— Deficiency in mobility due to an inadequate steering mechanism, which had a very high breakdown rate.
— Operations required generally specialized personnel: in the Wehrmacht an officer or Oberfeldwebel as tank commander, Unteroffiziers as gunner and driver.

Once the Germans no longer had any experienced tank crews, it was apparent that the Panthers were no longer employed operationally or were abandoned because of mechanical breakdowns.

In the spring of 1946 a Panther chassis was delivered to the Atélier de Construction de Moulineux (AMX) in Satory near Versailles. It was equipped with an LE 185 electro steering unit, and up until that time had been located at the Zahnradfabrik Friedrichshafen. The WaPrüf 6 contract for its development had been assigned back on 31 August 1942. 15,700 Reichsmarks had been set aside for its funding.

Exterior view of the ZF K 12 E 185 electro gearbox, rated at 650 hp.

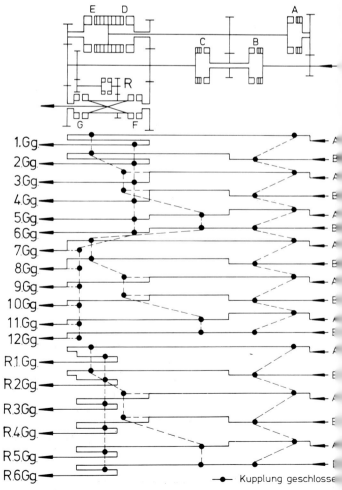

Shifting diagram for the K 12 E 185 electro gearbox.

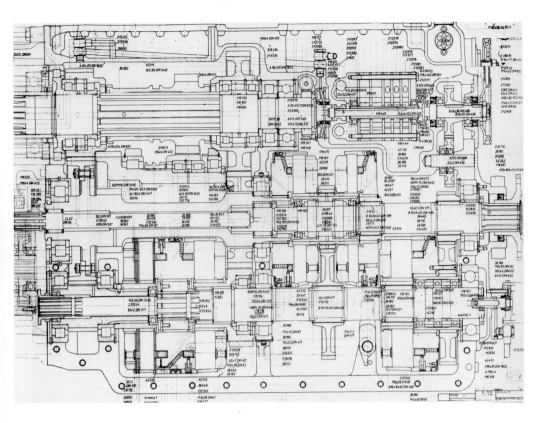

12 gear magnetic clutch gearbox, Model 12 E 185 for the Panther. Manufactured by Zahnradfabrik Friedrichshafen.

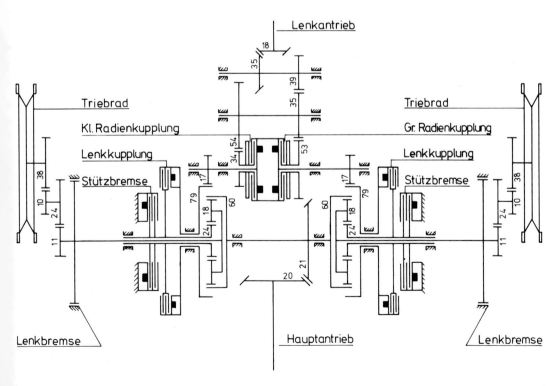

ZF LE-185 electro dual-radius steering unit for the Panther.

The development of night-vision devices (infra-red) for the Panther tank

The ever-increasing air superiority of the western Allies made daytime travel almost impossible for tank units. Great emphasis was placed on the development of an infra-red device to be used by the Panzertruppe for travelling and firing at night.

Beginning as early as 1936 the Heereswaffenamt, WaPrüf 8, in cooperation with the firm of AEG, began working on the problem of making infra-red scope technology applicable for use as driving, observation and targeting optics. In 1942 an infra-red gunsight was created for the 75mm Pak 40 self-propelled gun, which utilized an infra-red headlamp in addition to a viewer. The favorable results of troop testing with this device, which took place at the Panzertruppenschule Fallingbostel through 1943, led Guderian to demand that such an observation and sighting device be developed for the Panther as well. The results of this development, conducted under the direction of Ministerialrat Dr.-Ing. H. Gaertner of the Heereswaffenamt/WaPrüf 8, made use of a viewing device for the vehicle commander which could be removed during the day. An integral incorporation of the equipment into the overall design was not possible due to time constraints. The disadvantage of having the commander's head exposed outside the vehicle when using the equipment was negligible considering that the device was only to be used at night. Based on the military requirements, the device was expected to provide the following:
— 360-degree vision for the commander
— guidance of the driver by the commander
— targeting

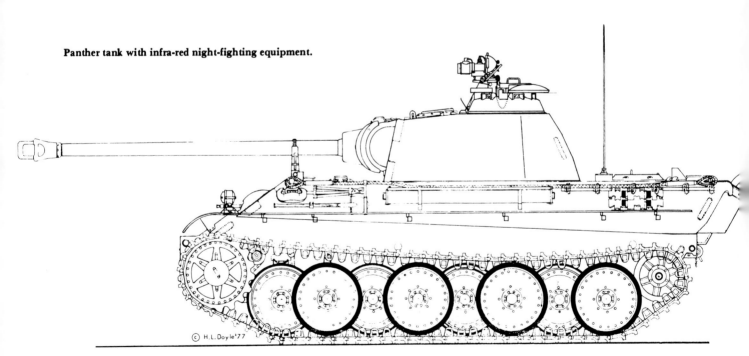

Panther tank with infra-red night-fighting equipment.

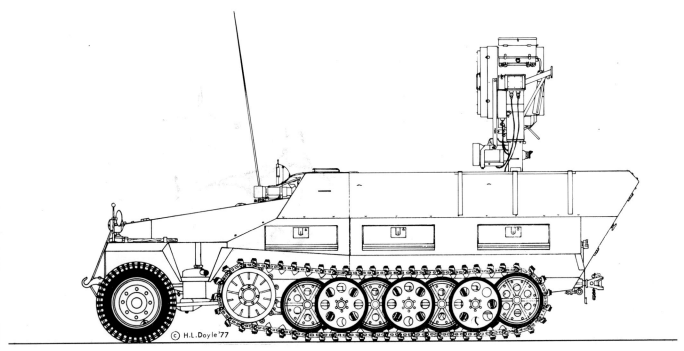
"Uhu" infra-red sighting vehicle (Sd.Kfz.251/20)

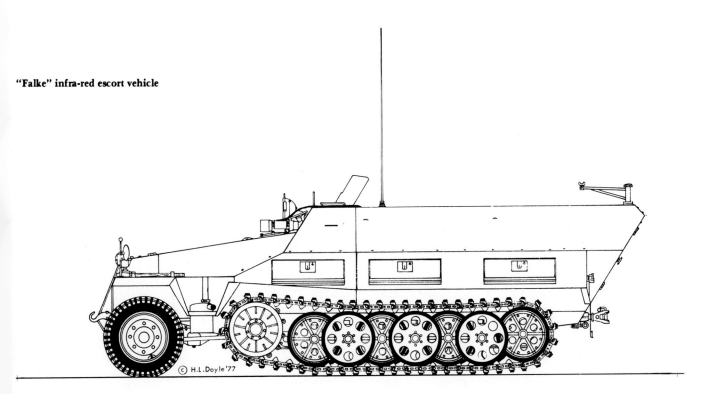
"Falke" infra-red escort vehicle

The device could be rotated a full 360 degrees by means of two support arms affixed to a ring around the commander's cupola. For guiding the driver the magnification was restricted to 1.2x, since with greater magnifications the landscape picture moved around too much when travelling. A diffusor in front of the headlamp increased the illuminated field from approximately 8 degrees to 30 degrees and gave the commander the ability to see ahead of the vehicle roughly 100 meters. The driver operated blind and relied on the directions from the commander. In spite of the low magnification targets could be acquired out to 400 meters. The target was aligned with the sighting marks within the viewer, the angle of elevation to the target being transferred via a steel band to an indicator inside the tank with an accuracy of +/- 1 degree. The gunner was then able to figure in the angle of elevation when adjusting the gun barrel. The alignment marks within the infra-red scope were adjustable in order to use the device with different types of ammunition. Electrical power was supplied by an auxiliary 400 watt generator. In addition, a 12 volt battery was built in. The headlamp had a diameter of 20 cm, with an output of 200 watts. Additional technical details of the system:

— Front lens unit (7 lens); f=9 cm; 1:1; field-of-view approximately 30 degrees
— Monocular magnifier; 5x enlargement
— Infra-red image-viewing tube Type 126 (triode) of AEG; operating voltage 18 kV

The inadequate range of the Panther's infra-red viewing system led to the development of an infra-red observation vehicle designed on the basis of the mittlerer Schützenpanzer armored personnel carrier (Sd.Kfz.251/20). It carried the designation "Uhu" (Owl). The vehicle held a 60 cm AA carbon-arc searchlight with infra-red filter as well as its associated batteries. An observer sitting beneath the system could traverse it a full 360 degrees; when the APC was on the move the equipment could be folded down into the vehicle. The overall magnification corresponded to that of the scissors sight (10x), giving an observation range out to 1500 meters. During tactical operations one observation vehicle would be assigned to a Panther unit comprising five tanks. The visibility range of the infra-red viewer mounted on the Panther tank was increased to approximately 700 meters through the improved capability of the Uhu's searchlight. Additional equipment on the Uhu included an infra-red night driving device, which gave the driver a speed of 40 to 50 km/h on good roads in complete darkness. The viewing lens for the infra-red viewscreen was of such a large diameter that the driver could use both eyes when looking into the device (which had a magnification of 1.2x). The

Panzerkampfwagen Panther Ausf. G with infra-red gunsight and headlamp mounted on the commander's cupola.

Platform for the infra-red gunsight and headlamp, front and rear views.

Sketch showing the details of the IR equipment.

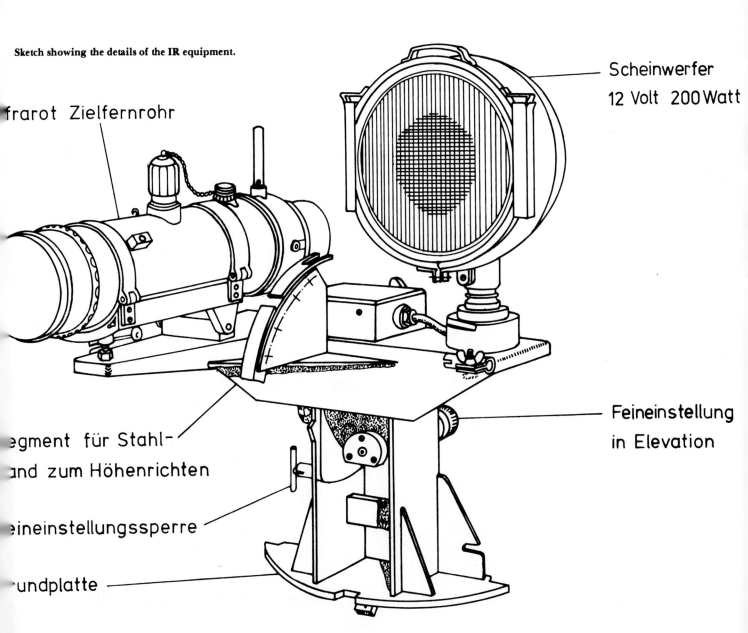

Infrarot Zielfernrohr

Segment für Stahl- und zum Höhenrichten

Feineinstellungssperre

Grundplatte

Scheinwerfer 12 Volt 200 Watt

Feineinstellung in Elevation

"Uhu" IR equipment mounted on a mittlerer Schützenpanzerwagen (Sd.Kfz.251/20) armored personnel carrier.

"Falke" infra-red driving equipment on a medium APC halftrack.

driving viewer was installed horizontally on the Uhu; it was also possible to mount the device vertically. There were plans to install an additional driving viewer installed vertically in the Panther, which would incorporate mirrors set at 45 degrees for both the eyepiece lens and the objective lens. Additional technical data for the Uhu observation vehicle:
— Infra-red searchlight, 60 cm diameter; 6 kW
— Infra-red driving headlamp, 20 cm diameter; 200 W
— Front lenspiece (4 lens) f=40 cm; 1:1.5; field-of-view approximately 10 degrees
— Monocular magnifier; 10x magnification
— Infra-red image viewing tube as found on the commander's equipment of the Panther tank

There was discussion of using two upward-folding incandescent lights, each of 60 cm diameter, in place of the single 6 kW carbon-arc searchlight. It had been demonstrated that these — when equipped with 500 watt bulbs — had the same range when the incandescent lights were powered with 28 V instead of 24 V.

An infra-red target scope was demanded for the Sturmgewehr 44 assault rifle when accompanying armor units during night operations with infra-red battlefield illumination. Panzergrenadiers equipped with these guns were to be transported in armored personnel carriers which carried the infra-red driving scope. The APC machine gun was also to have an infra-red gunsight. This Schützenpanzerwagen carried the designation "Falke" (Falcon).

Thus were the beginnings of a fully integrated night fighting capability, encompassing the commander's device for the Panther, the Uhu observation vehicle and the Falke escort vehicle. The commander's night sight reached a production output of roughly 100 pieces per month during the last few months of the war. In the end, when nearly 1000 devices were available, there was not only a shortage of tanks but also of fuel. Too little to be used in the first concentrated operation envisioned by Guderian. Other than isolated operations, this equipment was not used in great numbers by the front-line troops. Near the end of the war approximately 60 Uhu's were constructed out of a planned batch of 600. All parts had been sufficiently tested for mass production.

Thermal sensing devices were also developed during the Second World War, inspired by the efforts of Dr.-Ing Gaertner at the Zeiss firm. This was the "Donau 60" thermal detection device, which demonstrated its ability to sense a tank (a captured French vehicle) at a distance of 7.5 km. The device was never installed in a tank since the reception mirror had a diameter of 60 cm. These devices were, however, used on the Atlantic coast.

Panzerkampfwagen Panther II

In February 1943 the Heereswaffenamt, WaPrüf 6/III called for closer cooperation between Henschel and the Maschinenfabrik Augsburg-Nürnberg with the goal of achieving greater commonality between the Panther tank and the Tiger. In order to facilitate the supply of replacement parts, individual components were to be made interchangeable between the two vehicles. The improved vehicles would be designated Panther II and Tiger II. Although the Tiger II actually went into production toward the end of 1943 as the Ausführung B, only two prototypes of the Panther II were built by the end of the war. One of these vehicles, with the turret of the G model, survived and now is housed at the US armor training facility at Fort Knox in the state of Kentucky.

It was determined that the turret's greatest mass turning radius of the Panther II would not exceed a radius of 1570 mm. By reworking the corners the radius was brought down to 1565 mm, meaning that when the turret rotated a clearance of 15 to 20 mm was maintained from its edges to the driver's and radio operator's hatches. The total length from the center of the turret to its lower rear edge could only be 1240 mm so as to ensure access to the engine hatch in the hull roof. This requirement was met by changing the angle of the rear plate to 70 degrees instead of the originally planned 65 degrees.

In a meeting on the 5th of March 1943 it was established that the main drive and steering units (both with the Maybach OLVAR and the ZF synchro-mesh gearboxes) for the Panther II and Tiger II were to be built in such a manner that both could be installed independent of each other. Design plans for the steering mechanism and final drive would be completed by 20 March 1943.

On March 30th, 1943 the determination was made that the firm of Krupp would shortly begin production of the Panther II and would therefore be required to undertake the manufacture of the controlled differential discontinuous regenerative steering in cooperation with MIAG firm. MIAG would build the steering units while Krupp would make the final drives. Incidental to this arrangement, it was felt that production of the Panther II was not expected to begin prior to the end of 1944/beginning of 1945. Nevertheless, the Maschinenfabrik Augsburg-Nürnberg informed the Panzerkommission on

The Panther II prototype, which today can be seen at the tank training school in Fort Knox, USA, with a G model turret affixed.

The running gear of the Panther II made use of Tiger B parts (road wheel diameter 800 mm). The vehicle is being restored to driving condition at Fort Knox.

The Panther II had the single torsion bar suspension of the Tiger B.

Panther II running gear with all-steel road wheels.

Idler wheel and track arrangement of the Panther II was very similar to that of the Tiger B.

Panzerkampfwagen Panther II.

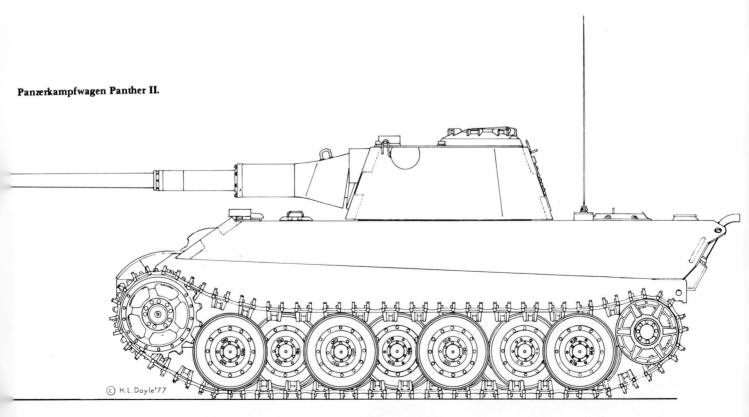

171

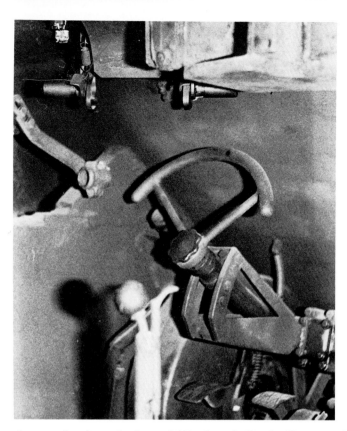 

As opposed to the production model Panthers, the Panther II prototype had a steering wheel in place of steering levers.

Aside from a 100 mm glacis armor thickness, the Panther II also made use of rotating periscopes for the driver and radio operator. The entry hatch arrangement of the G model was kept.

 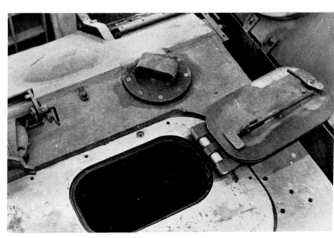

5 April 1943 that the initial Panther II prototype would be completed by mid August 1943. The all-up weight of the Panther II was given at over 50 tons. The armor of the glacis plate was increased to 100 mm, the sides were 60 mm and plates for the rear were 40 mm thick.

For reasons of material and cost conservation the closely arranged double torsion bars were dispensed with for the running gear; in their place the single torsion bars of the Tiger were to be utilized. The staggered running gear, which was planned to replace the previously used interleaved design, made use of all-steel resilient bogie wheels. The Tiger tracks would be used as well. It was also planned to install the Tiger's L 801 controlled regenerative steering mechanism and final drive. The steering mechanism was to be linked up with the strengthened ZF AK 7-200 all-synchro-mesh gearbox.

The smaller designed turret for the Panzerkampfwagen Panther, Ausf. F would also find use on the Panther II.

A meeting at Daimler-Benz on 27 February 1945 established that a design of the Panther turret with an 88mm gun would be accomplished through close co-operation between the firms of Daimler-Benz and Krupp. Initially, a mild steel prototype was to be manufactured by Daimler-Benz, Berlin-Marienfelde. The technical requirements were as follows:

— The gun was to be the 88mm KwK 43 of the Tiger II. Counter-recoil and recoil buffer cylinder were situated above with shell casing discharge cylinder in the center. The muzzle brake was dropped. The trunnion was moved to a different location.
— Turret front made of smooth armor plate with the smallest possible openings for the main gun and machine gun, i.e. the center of the trunnion must lie at the forward edge of the front plate.
— A rangefinder was planned. Efforts were made to make use of the tank rangefinder already developed by Zeiss, having a base of either 1.32 m or 1.65 m.
— A low-profile turret was considered advantageous.
— The free turret diameter was 1750 mm in order to provide the loader with the necessary freedom of movement.
— Ammunition was to be stored in the turret so as to be readily accessible.
— Commander's cupola and elevating/traversing gears from the Panther Ausf. F "Schmal-Turm" were to be utilized.
— A stabilized gunsight was to be considered.
— The turret rear plate was to be built slanted as opposed to Daimler-Benz' initial proposal.
— Krupp was given a contract to explore the feasibility of using gun stabilization in conjunction with the Tiger II turret. The gun would therefore need to be mounted at its center of mass.

Krupp suggested design testing the installation of a recoilless gun in the Tiger II turret. At a meeting on the 20th of February 1945 the designs of Daimler-Benz and Krupp were compared. Daimler-Benz had installed the 88mm gun, with its brake cylinder below, in a Panther turret with its operational circumference increased by 100 mm — necessitating a new turret bearing race. Krupp employed the KwK 43 already in use in an unmodified Panther turret, but with the trunnion pin moved. The Heereswaffenamt/WaPrüf 6 announced that the makeshift Krupp concept of installing the 88mm KwK 43 in Panther turrets was no longer of critical concern. A new design was to be considered with a view to the larger operating circumference, albeit with the standard 88mm KwK 43 (with altered trunnion pin) as proposed by Krupp. The development of the turret was to be accomplished by Daimler-Benz, while Krupp would be responsible for developing the gun. This developmental work was to continue despite the projected cessation of Panther production. The requirements laid down by the Heereswaffenamt/WaPrüf 6 for the turret were as follows:

— Elevation +15 to -8 degrees
— Installation of a rangefinder with 1.32/1.60 base
— Diameter of operational circumference at 1750 mm
— Trunnion pin in the front plate to provide the smallest possible opening for the optics and machine gun
— Ammunition readily accessible from within the turret

— Discharge cylinder instead of compressor
— Standard elevating and traversing gearing
— Sloped rear wall and previously used commander's cupola
— Smallest possible turret profile

This development was to be conducted under contract number SS 4911-0006-2937/44 Az 76 g 20 WaPrüf 6/11 dated 4 December 1944. The Krupp studies showed that the 88mm KwK 43 could indeed be installed in the turret if the trunnion pin on the cradle was shifted back 350 mm. In doing so, this moved the center of mass forward approximately 200 mm and increased the turret weight by 90 kg. Krupp calculated approximately 100 construction days for the manufacturing process.

The armor thicknesses for the turret remained unchanged. The "Saukopf" mantelet had a thickness of up to 150 mm. The design drawings for the Panther II turret, however, called for thinner armor plating, since the German steel industry was no longer able to produce a sufficient amount of armor plating in this thickness due to the advanced state of the war. A test model of the Panther II turret which today can be found in England shows a front plate of 125 mm, while the turret roof has a thickness of 40 mm.

During a conference in the Heereswaffenamt on 10 February 1944 the opinion was expressed that the Panther I no longer met the requirements in light of the experience gained on the Eastern Front. The Panther should be completely redesigned and, as already mentioned, receive the Tiger steering mechanism and final drive. The entire running gear and turret were to be changed. The new vehicle was given the designation Panther II. The weight of completed tank would be increased to over 50 tons. MAN was to make the plans available to the license builders in April/May 1944. In view of the new design, additional changes to the Panther I were to be avoided as far as possible. In spite of the bitter experience during the Panther I's production startup, the Panther II was to immediately enter into mass production without a pre-series batch.

According to a directive of the Rüstungsministerium, the firm of Demag, Düsseldorf-Benrath was to be immediately dropped from the manufacturing program of the Panther I (Demag had only built Bergepanthers). Demag and MIAG were to solely concern themselves with the Panther II. As far as is known, no further work was undertaken with regards to the series production of the Panther II.

With WaPrüf 6 contract number 006-5102/42 from 12 November 1942, ten test models of the Type 12 E 185 electro-gearboxes were ordered from the Zahnradfabrik Friedrichshafen for the Panther. The first of these gearboxes, having a total reduction ratio of 13.4 went into driving trials in June, 1944. It was a 12-gear magnetic clutch gearbox for 650 hp performance. Also in June of 1944 a 7-geared electro-gearbox for the Panther, designated E 130, was put through its paces. At the same time the AK 5-200 gearbox, reduced to five forward gears and with a reduction ratio of 9 was also tested; this was to be used in conjunction with the Deutz diesel engine. Additionally, these were also on the testing list:

— Gerät 544 — Panther with HL 234 engine in cooperation with MAN (same arrangement and displacement as HL 230 but with fuel injection — 850 hp at 3000 1/min — specific fuel consumption 220 g/hp/h)
— Gerät 545 — Panther with air-cooled Simmering diesel engine* (Type Sla. 16 — 16 cylinder X-shaped air-cooled four stroke diesel — 135 x 160 mm bore/stroke — displacement 36.5 liters — 720 hp at 2000 1/min)
— Gerät 546 — Panther with Argus diesel engine and lengthened hull (Type LD 220 — 16 cylinder H-shaped air-cooled four stroke diesel — 135 x 165 bore/stroke — displacement 37.8 liters — 700 hp at 2200 1/min)
— Gerät 570 Panzerwagen 605/5, of which no other information is known.

Also worth mentioning here is a recommendation by the Ministerium für Rüstung und Kriegsproduktion from 1942 calling for the Panther to be built as a reconnaissance vehicle along with the planned Leopard combat reconnaissance vehicle. The turret under consideration had originally been developed by Daimler-Benz for the 8-wheeled Panzerspähwagen ARK. With somewhat increased armor, it was to be used on both the Leopard and the Panther reconnaissance vehicles.

---

* Dr.-Ing. h.c. F. Porsche's company had ceased its cooperative effort on the 16-cylinder X-form tank engine with Simmering-Graz-Pauker, Vienna, in August 1943.

Performance graph charts for the tank engines in development (data for performance, r.p.m. and fuel consumption)

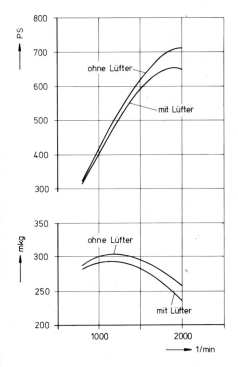

— Deutz two-stroke T 8 M 118 diesel engine

— Maybach four-stroke HL 234 internal combustion engine with fuel injection

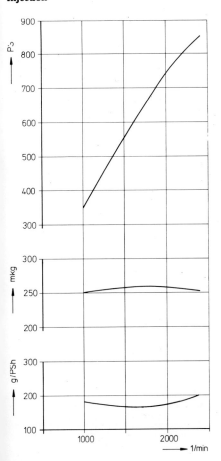

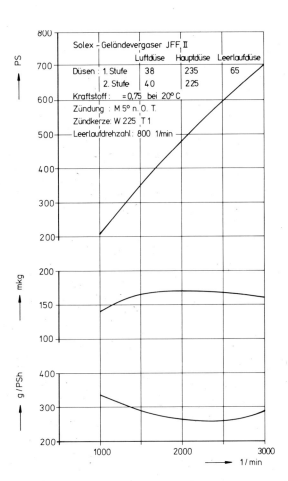

— Maybach four-stroke HL 230 internal combustion engine

— Daimler-Benz four-stroke MB 507 diesel engine

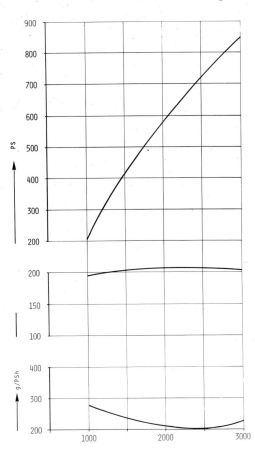

The contract for the vehicle's gun was awarded in January 1942 to the firms of Krupp and Rheinmetall. The requirements dictated the use of the 50mm KwK 39 L/60 barrel, but having a smaller design. A projected delivery was foreseen in the October/November 1942 timeframe. Neither project was pursued further.

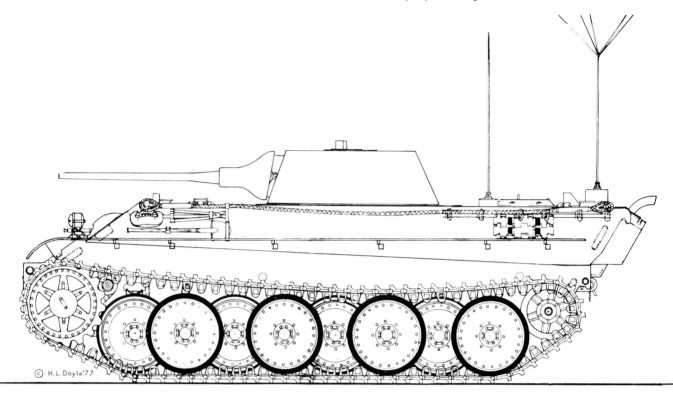

Panther battlefield reconnaissance vehicle (proposal)

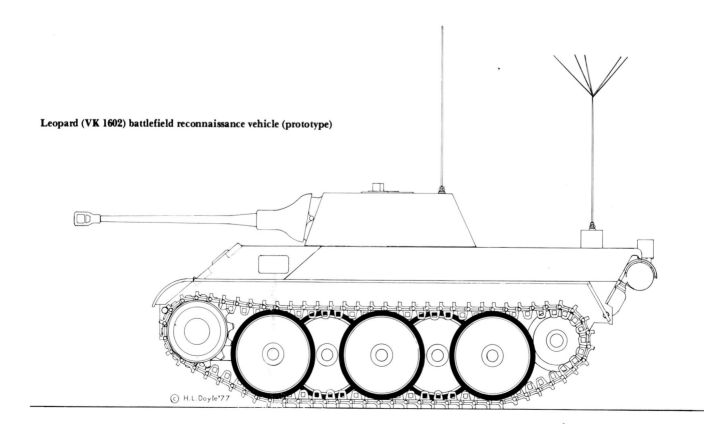

Leopard (VK 1602) battlefield reconnaissance vehicle (prototype)

Panzerbefehlswagen Panther and Panzerbeobachtungswagen Panther vehicles

A certain number of Panther tanks were delivered as armored command vehicles (Panzerbefehlswagen). These were equipped with two different radio sets:
— Sd.Kfz.267 with Fu 5 and Fu 8 radio equipment
— Sd.Kfz.268 with Fu 5 and Fu 7 radio equipment

The Panzerbefehlswagen Panther was created by modifying the tank with communications equipment. The modified vehicle dispensed with:
— the machine gun in the gun mantelet; a machine gun accessories box with mount; a machine gun toolbox with mount; as well as the machine gun barrel sleeve and retainer.
— 15 rounds of 75mm ammunition from the D and A models, 12 from the G model, including the ammunition racks (to the right behind the gunner in the hull)
— the loader's auxiliary turret traverse mechanism (Ausf. A)

The following communications equipment was installed:
— an on-board intercom system for command vehicles
— an Fu 5 radio set (10 Watt transmitter and VHF receiver) in the turret; either an Fu 7 radio set (20 Watt transmitter and VHF receiver) or an Fu 8 radio set (30 Watt transmitter and a medium wave receiver) within the hull, including their mounts
— a GG 400 generator
— a radio accessories box
— an antenna feed with a star antenna for 30 Watts (Fu 8); a 1.4 m rod antenna for 20 Watts (Fu 7) and a 2 m rod antenna for 10 Watts (Fu 5)
— an armored casing with antenna lead and star antenna in the center of the rear deck
— receptacles for collapsible radio masts and antenna rods on the exterior of the rear deck

The armament of the command tank consisted of the 75mm KwK 42 L/70 in the gun mantelet, an MG 34 machine gun in a ball mount (from Ausf. A) in the glacis, an MG 34 for anti-aircraft defense on the turret and an MP 40 machine pistol.

The crew comprised the commander, the communications officer (gunner), radioman 1 (loader), radioman 2 (radio operator) and driver.

64 rounds of 75mm ammunition, along with 34 bags each having 150 rounds for the machine guns were available. An additional 12 flare rounds, 12 smoke rounds with Zündschraube C 43 fuse and Wurfladung 1 propellant charge for the close-combat defense weapon as well as 20 326 Lp HE rounds were carried.

The star antenna was mounted on the right of the rear deck armor, the 1.4 meter rod antenna on the left of the rear deck and the 2 meter rod antenna on the turret roof to the right and behind the commander's cupola. These three antenna were anchored in rubber antenna feet. These were flexible, so that the antennae would give all the way to the horizontal when brushing against obstacles and right themselves automatically. The antenna rods were hollow, tapered and made of sheet steel; they were held in their feet by terminal screws.

In the hull, the mounts for the Fu 7 and Fu 8 radio sets were arranged one above the other to the left of the second radioman's seat. The retainers for the Fu 5 radio set were next to each other on the inner turret side wall to the right

Both the A and G models were supplied as command vehicles (a model A of the Panzer-Division Grossdeutschland)

Panther Panzerbefehlswagen (armored command vehicle). Externally, it differed only in the additional antenna equipment. Photos of an Ausf. D.

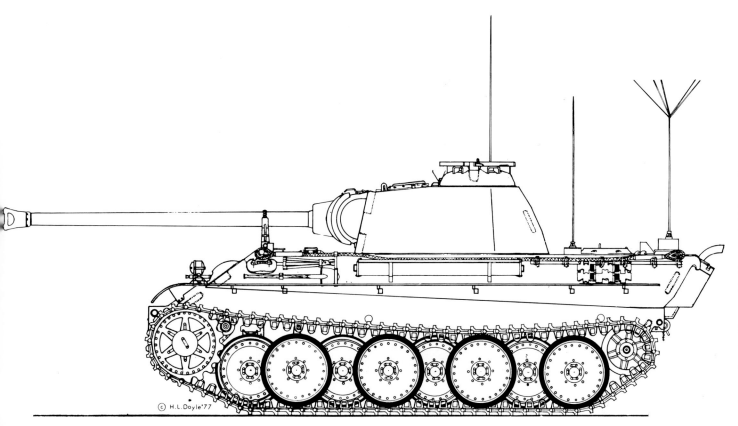

Panzerbefehlswagen Panther (Sd.Kfz.267)

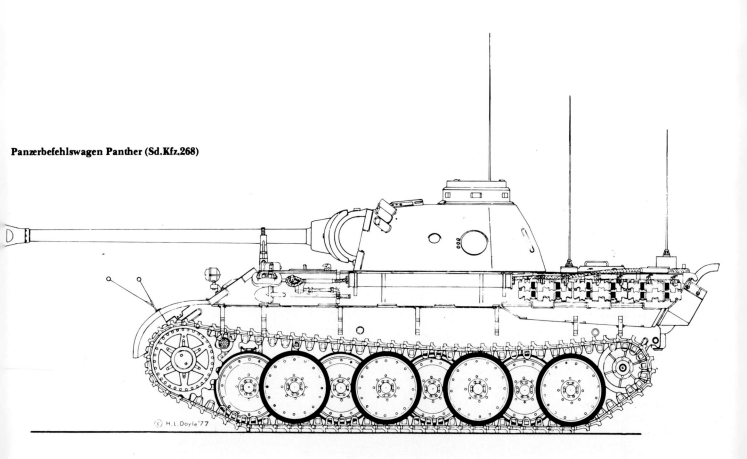

Panzerbefehlswagen Panther (Sd.Kfz.268)

Turret of a command vehicle from the left and right sides

Anti-aircraft machine gun mount

Turret of the command vehicle from the front

Seitenrichtungsanzeiger 7,5-cm-Kwk 42 (L/70)   Lüfter   Absaugleitung

Turm-
schwenk-
werk

Höhen-
richt-
handrad

Kommandanten-
sitz

Geschützausgleicher   Höhenrichtmaschine

Turret interior of the Panther Panzerbefehlswagen. On the turret side wall to the right the mounts for additional radio equipment can be discerned.

Lagerung für T.Z.F. 12a   Leuchte

Rohrzurrung

Rohrausblase-
vorrichtung

Abweiser

T.Z.F. 12a   Luftausgleicher

Seitenrichtungsanzeiger   Rohrzurrung 7,5-cm-Kwk 42 (L/70) Lüfter Winkelspiegellagerung

Absaugeleitung

Dichtung
für Blende

Rohrbremse

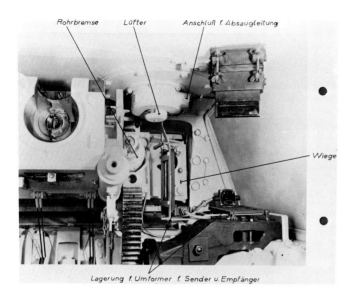

Mountings for the sender and receiver transformers.

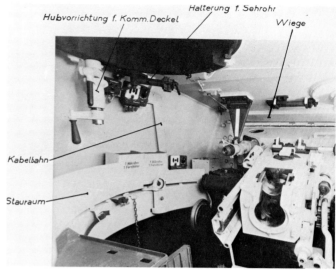

Mountings for the senders and receivers on the right side of the turret.

Right side of the turret basket with the secondary radio operator's seat.

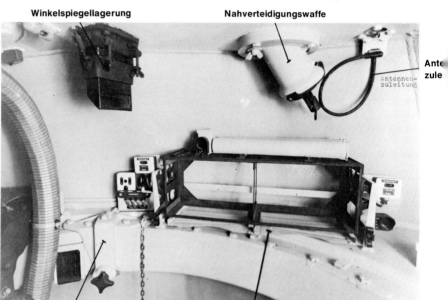

View of the commander's seat (minus cushion).

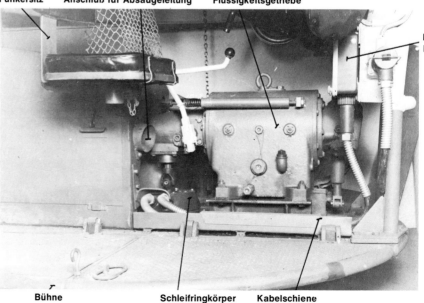

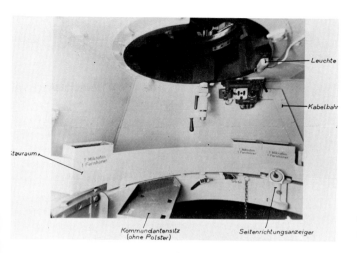
Side and rear walls of turret with commander's cupola.

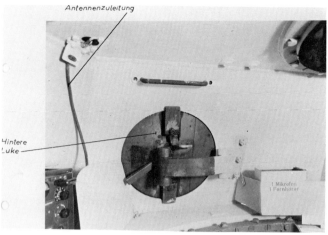
Rear escape hatch. The line feeds for the additional antennae in the turret roof are also visible.

of the first radioman.

The mounts for the transmitter and receiver were cushioned in a frame by rubber supports, the frames being anchored in the hull to the forward hull roof plate and in the turret to the inner turret wall. The transformers were arranged in a framework above the elevating mechanism housing.

Rheinmetall-Borsig, in conjunction with the firm of Anschütz in Kiel, developed a Panzerbeobachtungswagen Panther, or Panther observation vehicle, for use as fire direction vehicle by the armored artillery. The vehicle, based on the Panther Ausf. D, did not utilize the 75mm KwK 42 L/70, but had a dummy gun with an MG 34 in a ball mount on the front plate of the turret. The turret was traversable through a full 360 degrees and was equipped on both sides with smoke dischargers. The crew consisted of four — commander, observer, radioman and driver. The dummy gun was made of sheet steel and mounted to the front of the turret. The gun mantelet covered just over a third of the turret frontpiece. A Kugelzielfernrohr 2 (KZF 2) with 1.8x magnification served as the sighting device. The machine gun could be swiveled 5 degrees to either

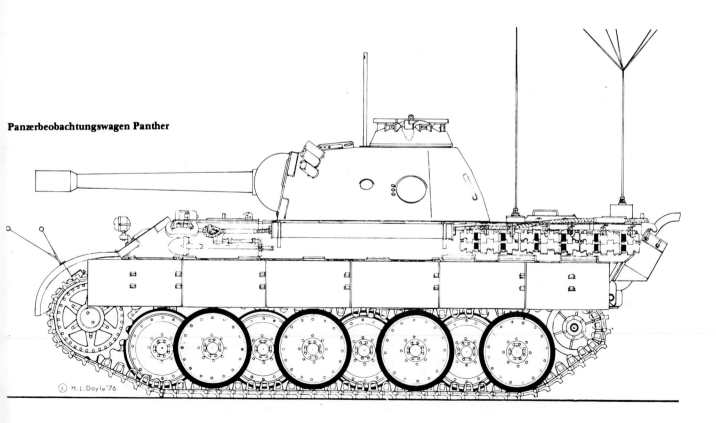
Panzerbeobachtungswagen Panther

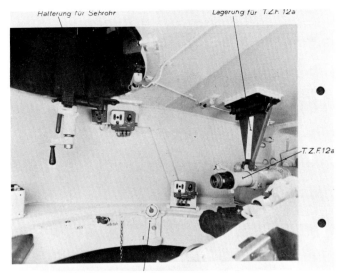

Gunner's station, in this case for the communications officer. The vehicle still maintained full combat capability.

Main fuse box.

Turret roof of the Panther command tank.

Center: Fu 5 radio set on the right turret side wall for the primary radio operator.

Below: Panzerbeobachtungswagen (armored mobile observation vehicle) Panther with dummy gun.

side, and had an elevation range of -10 to +15 degrees.

Additionally, the vehicle also carried an MP 40 machine pistol and a flare pistol. In addition the following optical instruments were employed: a rangefinder with a 1.32 meter base, two TBF 2 turret-mounted observation scopes (one as a replacement), two TSR 1 commander's observation devices and a scissors periscope.

The rangefinder with a 1.32 meter base and 5x magnification was set immediately behind the turret front plate. The vision ports on each side could be closed off by hinged cover plates from within the turret. The scales on the rangefinder could be illuminated at night. The firm of Zeiss was the manufacturer for the rangefinder. The turret-mounted observation scope was fixed in a ball mount in the center of the turret roof and could be extended upward approximately 45 cm. When not in use the turret opening was closed off with a cover plate. Its range of view was 360 degrees. By utilizing two adjusting screws the observation scope could be tilted 10 degrees.

The TSR 1 commander's observation device or the scissors periscope could be affixed to an adjustable bracket ahead of the commander's cupola.

Both the commander and the observer each had a turret direction indicator for their use. The gauge for the commander was the same as that found in a normal tank; for the observer a gauge with two scales was fixed to the forward turret wall below the center of the rangefinder. The firm of Anschütz produced a so-called "Blockstelle 0" (automated plotting board) for the Panzerbeobachtungswagen Panther, which was set in a shockproof mount and installed in the turret ahead of the commander's cupola. With this artillery sighting device it was possible to pass firing orders to the artillery even without map materials. Equipment was called for which permitted monitoring fired shells over greater distances. A star antenna was utilized for the additional radio equipment. The radiotelephone had an auxiliary loudspeaker. The Panzerbeobachtungswagen Panther did not see operational service other than for testing.

Jagdpanther

On 6 January 1942 the firm of Krupp turned over to the Heereswaffenamt construction drawings for the Panzerselbstfahrlafette IVc, which was designed to give mobility to the 88mm Pak 43 L/71. The fully tracked chassis previously used had proven inadequate. Whereas the chassis of the Panzerkampfwagen III and IV were too weak in addition to having an open fighting compart-

Wooden model of the Jagdpanther, produced by Krupp.

ment, the tank destroyer based on the Porsche Tiger was too complicated and cumbersome. The Pz.Sfl. IVc 2 developed by Krupp was to be equipped with a vertical Deutz radial diesel engine*; alternatively, the installation of the Maybach HL 90 engine would be studied. The Heereswaffenamt/WaPrüf 6 was in agreement with the concept of the vehicle. Krupp assented to developing a further design with regards to the following points:
— Installation of the Maybach HL 90 engine (the Deutz diesel engine was to be tested in a prototype IVc)
— 30 ton total vehicle weight
— Armor: 80 mm front, 40 mm sides
— 40 km/h maximum speed
— Gun depression: 10 degrees (previously 8 degrees)
— Alternative installation of leaf spring running gear in place of the torsion bar suspension
— Gun travel lock on base plate

The vehicle was designed for accompanying infantry; initially a maximum speed of 15 to 20 km/h was planned. When using a leaf spring suspension in place of the torsion bars the muzzle height and overall height was reduced by roughly 100 mm. On 2 April 1942 the wooden mockup of the vehicle was seen by the Heereswaffenamt, WaPrüf 6. Based on this, the following recommendations were made:
— Increasing the ground clearance from 400 to 450 mm

---

* Type T8 X113 — water-cooled 2 stroke diesel engine — 8-cylinder radial, 130x130 bore/stroke — 13.8 liter swept volume, 400 hp at 2500 1/min.

One of the initial Jagdpanther vehicles as evidenced by the double driver vision ports and the pistol port on the side of the superstructure.

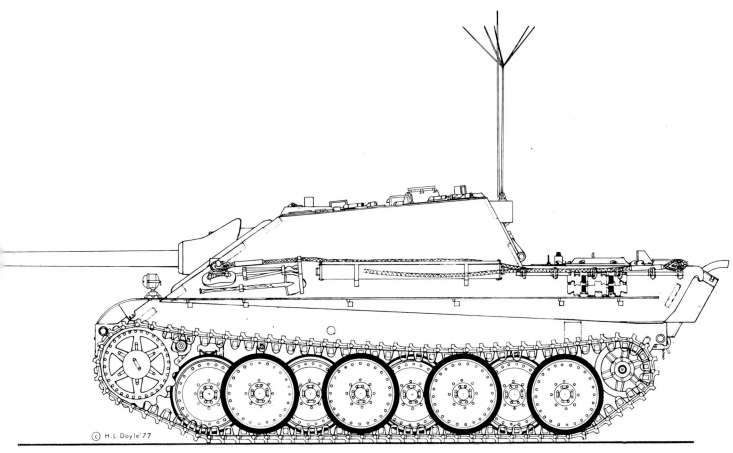

Jagdpanther (Sd.Kfz.173) (initial design) seen here as a commander's vehicle.

— Reducing the barrel depression from 10 to 8 degrees, in order not to increase the overall height of the vehicle with increased ground clearance.
— Using the permitted vehicle width of 3140 mm, studies were to be undertaken for reducing the ground pressure by utilizing currently available wider tracks.
— Viewing to the sides and rear by using periscopes and viewports
— The driver's periscope was to be traversable
— The installation of an armored periscope commander's cupola was to be studied, along with the installation of an AA gun in a rotating turret.
— A targeting mechanism for indirect sighting was dropped

Jagdpanther as a testbed at the Verskraft in Kummersdorf (chassis number V 102).

A wooden mockup was used to explore the arrangement of the gun and ammunition.

The Jagdpanther, with its superior gun, proved to be an effective countermeasure against all enemy tanks.

The Fried-Krupp Grusonwerk AG announced on 17 June 1942 that the production of three complete prototype armored self-propelled vehicles would interfere with Panzer IV output and proposed that the production be handed over to the firm of Deutsche Stahlindustrie, Mühlheim (Ruhr). The contract for the three improved test vehicles, now designated Pz.Sfl. IVd was assigned to the AK (Artillery Construction) department of Krupp on 9 June 1942 under contract number SS-006-4723/42. On 29 June 1942 this department requested that the contract be assigned directly to Krupp-Gruson. On August 3rd, 1942 the Heereswaffenamt/WaPrüf 6/IIe announced that the Panther chassis (available beginning August 1942) would be used for the heavy tank destroyer (88mm L/71). The corresponding parts for the Aufklärungspanzer Leopard reconnaissance vehicle only became available after June of 1943. The Heereswaffenamt/WaPrüf 6/III had already planned Panther chassis for the three Jagdpanzer prototypes. The Heereswaffenamt took the occasion of a conference at Krupp in Essen on 9 August 1942 to

The outstanding ballistic shape of the vehicle is readily apparent.

A Jagdpanther during a demonstration. Two pistol ports can be seen in the right side. Behind the Jagdpanther is a Tiger B with the production turret.

Jagdpanther from above with some hatch covers removed.

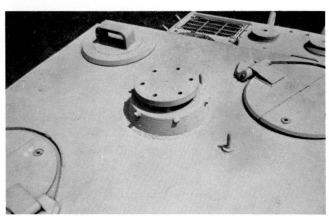

Ventilation cover, entry hatch and rotating periscope for the loader in the superstructure roof.

"Kidney" hatch for the gunner's observation equipment

Right and left sides of a fully-equipped Jagdpanther.

announce that the Pz.Sfl. IVd would be built as a Panzerjäger (tank destroyer) as part of the continuing development of the assault gun series. A monthly output of 60 units was projected from Krupp-Gruson in place of the planned Panther production. In addition, a 75mm Sturmgeschütz assault gun was under consideration, but the design of this had not yet been established.

Since the Panther chassis required a complete redesign for use as a tank destroyer, Krupp was only able to provide the necessary plans in January of 1943. There was pressing interest in having the first chassis delivered by June 1943 with series production beginning in July 1943. Krupp promised a 1:10 scale wooden model by the end of September and a 1:1 scale mockup by 10 November 1942. With minor changes it was possible to make use of the Panther chassis. The gun to be installed was basically the 88mm KwK L/71. The Heereswaffenamt/WaPrüf 6 sent an example of a 1.5 m long periscope for purposes of observation (due to striations and residue development). Fundamental data for the vehicle was established as follows: combat weight of approximately 35 tons; main armament 88mm KwK L/71, an MG 42, two machine pistols, front armor 80 mm at 90 degrees, 50 mm at 30 degrees, side armor 40 mm at 50 degrees; 60 rounds ammunition; muzzle height 1850 mm; overall height 2400 mm; total length 9000 mm; movement of gun +14 degrees elevation, side +/- 14 degrees.

As a result of a meeting on 15 September 1942 in the Reichsministerium für Rüstung und Kriegsproduktion, the developmental work which had been performed by Krupp up to that time was concluded with the utmost appreciation; further development of the 88mm Panzerjäger Panther was to be placed in the hands of Daimler-Benz, Berlin Marienfelde, since the production of the vehicle was already scheduled to begin there in the summer of 1943. By assimilating both design and manufacture at Daimler-Benz, it was expected that the production run would be simplified and at the same time it would also free up a portion of the design bureau at Krupp for special projects.

Krupp was requested to conclude the design at a particular point and in agreement with Daimler-Benz, carry out the transfer of labor in such a manner that the series maturation could be accomplished at Daimler-Benz. The 1:1 scale wooden mockup completed by Krupp was inspected on 16 November 1942.

On 5 January design and manufacturing questions regarding the 88mm Sturmgeschütz (Panther) were addressed at Daimler-Benz in Berlin-Marienfelde. As a

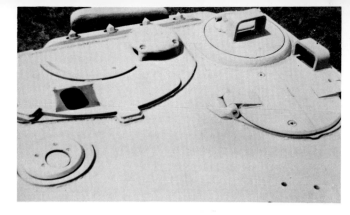

Below left can be seen the close-combat defense weapon, while the observation periscope and hatch for the loader are on the right.

Ball mount for the radio operator in the frontal plate of the Jagdpanther.

The large hatch at the rear of the superstructure served for working on the gun. The smaller hatch to the right of it was for dispensing empty cartridges.

Jagdpanthers during operations in Normandy. Notice the covered outer driver's vision block.

result of this, it was determined to produce the gun collar of the current design with its continuous frontplate (35 degree slope) only in cast steel. The Brandenburger Eisenwerke was requested to work towards a molybdenum-free cast. The armor thicknesses were to be as follows: upper front 100 mm; lower front 60 mm; deck, rear and bottom plate 30 mm. The gun was to be offset to the right as far as the space constraints for the driver dictated. The gun collar was installable and removable from the front, permitting the removal of the gun carriage from the front once the barrel had been removed through the rear. With the transmission able to be removed through the front it was possible to have the superstructure roof welded solid. The attempt was made to position the rear superstructure wall vertically, with a hinged hatch-plate. Daimler-Benz developed a travel lock holding the

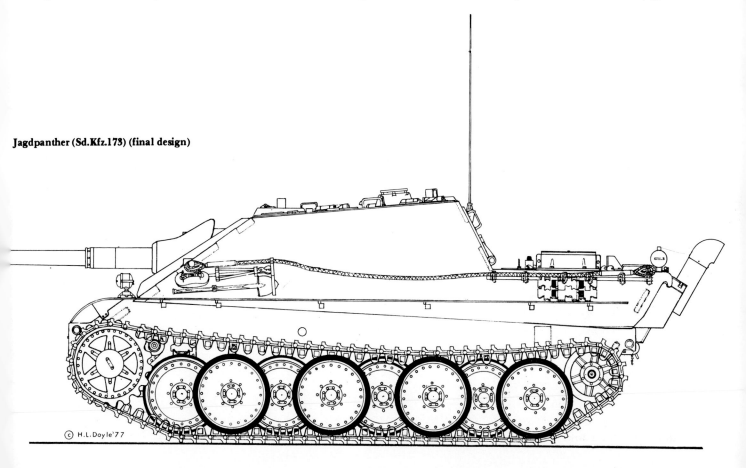

Jagdpanther (Sd.Kfz.173) (final design)

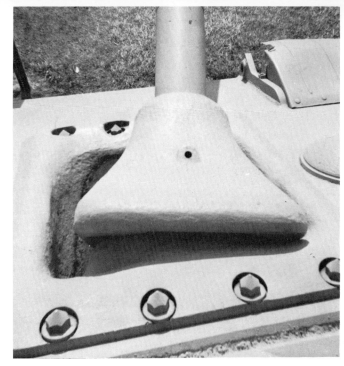

Instead of the welded mantelet the final production examples of the Jagdpanther were given a bolted collarpiece which covered the gun opening.

Later models had only one opening for the driver's optics and two-piece barrels for the 88mm Pak.

Along with two-piece barrels, single-piece barrels were also used on the Jagdpanther. This one also carries the bolted collar.

The running gear was not changed from the Panther tank.

Size comparison between the Jagdpanther and the Jagdpanzer Hetzer.

Comparison of the tank destroyers used by the German Wehrmacht.

Jagdpanzer 38 "Hetzer"

Sturmgeschütz III

Jagdpanzer IV

Jagdpanther

Jagdtiger

barrel at 0 degrees elevation. In place of a driver's vision port two periscopes were installed. Pistol ports were planned to provide the driver with side views. Daimler-Benz studied the possibility of installing scissors periscope for the commander in the planned twin machine gun cupola design. The first armor vehicle was expected by mid 1943 with production of series vehicles commencing at the beginning of December 1943.

The chassis of the Panzerkampfwagen Panther, Ausf. G was also utilized for the schwere Sturmgeschütz 88mm (Panther) design. Up to this point, the vehicle was based on the Panther II of Daimler-Benz and was now required to be converted over to the basis of the Panther I, given the new situation and taking into account simplification of production. As a result of extending the upper sideplates, the weight was increased by 800 kg, which was only mitigated by cutting off 250 kg by reducing the thickness of the lower front plate and bottom plate armor. A weight increase of 550 kg would have to be taken into account. An additional 50 kg could be saved by modifying the vehicle's fuel tanks.

On 24 May 1943 engineers from the Mühlenbau- und Industrie AG (MIAG), Braunschweig discussed details of the Panther's main gun arrangement at Krupp. MIAG had assumed the construction of the heavy assault gun based of the Panther from Daimler-Benz. Daimler-Benz was still in charge of gun installation and removal, but in close cooperation with MIAG. MIAG reported that the superstructure deck would only be 16 mm thick and recommended re-exploring the proposal of mounting the barrel travel lock to the superstructure roof. There were attempts made to produce this lock affixed to the hull bottom plate. For reasons dictated by space the gun was required to be shifted 50 mm to the right. In doing so, the position indicator on the gun collided with the shock absorber when the gun was at its highest elevation and traverse of 12 degrees. MIAG recommended reducing the lateral movement from +/-12 degrees = 24 degrees to -12 degrees and +10 degrees = 22 degrees.

On 2 June 1943 it was determined to dispense with the concept of indirect target acquisition; the previous deck cutouts for the sighting arrangement would now have to be changed. There were plans to equip the vehicle with a gas protection system; sealing of the gun collar was undertaken by Daimler-Benz. The company recommended shortening the cradle armor by 30 mm, which, however, was never implemented.

On 9 June 1943 the following details were established for the 88mm Panzerjäger 43/3 (L/71) auf Fahrgestell

Panther I:
— Designation of vehicle: 88mm Panzerjäger 43/3 (L/71) Panther
— Chassis: Panther I
— Armament:
a) 88mm Pak 43/3 (L/71) — the gun, including carriage, elevating and traversing mechanisms was developed by Krupp
b) four mounts and five firing ports for the MP 40
c) smoke grenade discharger — plans and test model for installation in the wooden mockup would be sent through WaPrüf 6
d) An MG 42 for engaging ground targets, if possible with 360 degree arc of fire
— Blowout system for barrel is to be installed; the design plans would be sent via the Heereswaffenamt/WaPrüf 4 after completion
— Crew: 6 men (1 commander, 1 driver, 1 gunner, 2 loaders, 1 radio operator)
— Rounds of ammunition: 50 for the Pak; 30 for the smoke discharger; 600 for the machine gun; 760 for the machine pistol
— Field of fire: elevation +14 degrees -8 degrees, lateral +/-12 degrees
— Armor: front — upper 80 mm, lower 50 mm; rear — 40 mm; sides-upper 50 mm, lower 40 mm; roof and bottom — 16 mm
— Sighting equipment: initially Zieleinrichtung 37 with Sfl Zf 1a. The installation of the planned replacement for the Zieleinrichtung 37, the W.Z.F. 1 should be taken into account. The cutout above the gunner is to be covered by a sliding hatch controlled from the carriage. Fire direction equipment for relaying information from the commander to the gunner is foreseen
— Observation equipment: SF 14 Z (Sfl); periscopes for the commander (3 pair); driver (2 pair); loader each 1 (2

The Jagdpanther of the author, which was knocked out on 16 April 1945 in the Gifhorn area.

The Jagdpanther was built in this form until the war's end.

pair), one pair being traversable; reserve (6 pair)
— Radio equipment and on-board intercom will be based on further information from the Heereswafenamt/WaPrüf 7
— Entry hatches are planned for above the commander, the left loader and in the rear wall of the superstructure, the latter being the maintenance hatch for the main gun
— For communication between the commander and driver a mechanical device is to be used in addition to the intercom (a speaking tube or something similar to a machine telegraph as developed by the firm of Telekin)
— Incorporation of an infra-red sighting device and IR

target illuminating headlamp is planned. Drawings will be supplied
— An air filtration system is to be installed
— General drawings including dimensions of armor and weight calculations are to be turned over to the Heereswaffenamt/WaPrüf 6 Pz IV
— The 1:1 scale wooden model, currently at Daimler-Benz, is to be assembled at MIAG no later than 15 June 1943
— The contract awarded to Brandenburger Eisenwerke for two models to be used in testing armor penetration is to be fulfilled by 15 June and 1 July 1943, respectively.

The 1:1 scale wooden mockup was demonstrated before Hitler on 20 October 1943. He was presented with the first photos of the Panzerjäger auf Panther-Fahrgestell on 15 November 1943. Hitler considered this design as an excellent solution and called it an "armored casemate." He once again emphasized the significance of producing assault guns and tank destroyers, which at this juncture in the defense strategy he valued more highly than the related tank.

On 17 December 1943 the first demonstration of the Jagdpanther Panther took place. An enclosed, ballistically well-formed structure minus turret, which held an 88mm Pak 43/3 (L/71) was set onto an unmodified Panther chassis. The gun opening, which had been kept relatively small, was covered by a collar (welded on the first versions). The lateral movement was +/-11 degrees = 22 degrees, the elevation range was -8 to +14 degrees. Armor thickness was 80 mm upper front, side 45 mm and rear 40 mm. The all-up combat weight with a crew of five and 60 rounds of ammunition amounted to 45.5 tons. Chassis numbers ranged from 300001 to 300392. The manufacture of the armored superstructure was turned over to the Brandenburgische Eisenwerk Kirchmörser in Brandenburg/Havel, whereas the main gun was assigned to the Dortmund-Hoerder Hüttenverein, Werk Lippstadt. The radio operator, sitting on the front right, was provided with an MG 34 installed in a ball mount. Hitler took the occasion of the demonstration on 17 December to re-emphasize the importance of these kinds of vehicles. He expressed the view that, for example, the Panzerjäger

Front view of the Jagdpanther, early model (left) and final design (right)

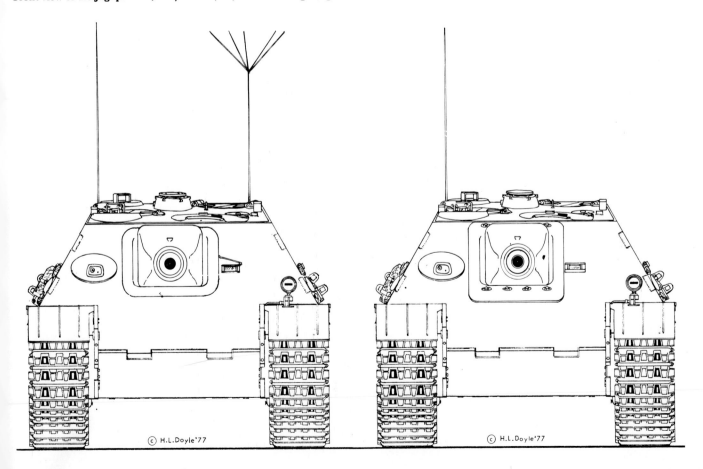

Panther would eventually prove to be superior in many cases to the Tiger II (equipped with the same gun) and pointed to the lower manufacturing costs, the lower weight, the reduced susceptibility to shell damage and the greater mobility of the vehicle with the same engine thanks to its lower weight.

Later versions of the Panzerjäger Panther had a strengthened bolt-on collar for the gun opening. With a few minor changes the 88mm Pak L/71 was designated 43/4.

There were two models envisioned for the Panzerjäger Panther:
— Sturmgeschütz Panther für 88mm StuK 43 (Sd.Kfz.172)
— Panzerjäger Panther für 88mm Pak 43/3 or Pak 43/4 (Sf) (Sd.Kfz.173)

A directive from the Führer on 27 February 1944 determined that the schwerer Panzerjäger auf Panther I-Fahrgestell would now be designated "Jagdpanther" (Hunting Panther).

Aside from the AK 7-200 synchro-mesh gearbox from the firm of ZF, there was also studies undertaken investigating the possibility of using the semi-automatic Maybach OLVAR gearbox of the Tiger. Production of the Jagdpanther began in January 1944 at MIAG where it was produced until the war's end. In November of 1944 the Maschinenfabrik Niedersachsen-Hannover (MNH), and starting in December of 1944 the MBA (Maschinenbau und Bahnbedarf) also assumed production of the Jagdpanther. All three manufacturers produced the following numbers of vehicles: MIAG 270, MNH 112, MBA 37.

A total of 415 Jagdpanthers were built. They were used by several Heeres-Panzerjägerabteilungen up until the end of the war. On 1 February 1945 the Jagdpanther production run was projected to stop at 450 vehicles manufactured.

The author was the company commander of a Jagdpanther unit near the war's end and was knocked out in a Jagdpanther on 16 April 1945 near Gifhorn.

The Fried-Krupp AG also became involved in up-gunning the main gun of the Jagdpanther. Drawing number Hln-E 143 from 17 November 1944 shows the installation of a 128mm Pak 80 L/55 in a Panzerjäger Panther.

During a conference in Kummersdorf on 25 October 1944 criticism was voiced regarding the Panther, which apparently had a rate of fire (Feuerfolge) significantly less than that of the American Sherman tank. The Heereswaffenamt/WaPrüf 6 intended to conduct firing trials between the Tiger II and the Sherman and, at the same time, inform on the time gained through the use of the Pak 43/3 multiple loader. The Heereswaffenamt/WaPrüf 4 announced that the first test model of the Pak 43/3 multiple loader was to be produced by 1 December 1944.

On 8 February 1945 Krupp made the announcement that the two multiple loader drums had been completed. The drums and an 88mm Pak modified by Dortmund-Hoerder Hüttenverein to accept a multi-loader drum were transferred to the Hillersleben Firing Range. A second gun was to be modified for installing a Pak 43/3 with drum in the Jagdpanther; in addition to the changes carried out on the first gun, this gun was to have its trunnion pin shifted back by approximately 400 mm. A meeting on 20 February 1945 at the Krupp firm in Essen called for an automatic multi-loading system for tanks, self-propelled vehicles and ammunition carriers. Studies were conducted with 88mm and 128mm calibers involving a new project for multiple loaders providing 3 to 5 rounds of ammunition; these were to be developed in conjunction with multiple loader for the 88mm Pak 43/3, already expected to enter production.

In October of 1944 a barrel blowout device was explored for the Jagdpanther as well. The test device, with its cylinder resting between the counter-recoil and the recoil brake, was installed in a Jagdpanther at the Verskraft. There were no difficulties, the elevation and traverse ranges not being affected. After each shot there followed a smooth discharge of the casing from the barrel. The control of the discharge initiation as the breech was opened was properly adjusted. Gasses from the cartridge casings were so thoroughly removed by the deck vents that there was only minimal smoke buildup within the fighting compartment. As a result of the successful firing trials it was proposed that a comprehensive incorporation of the Skoda shell casing discharger (using piston compression) be undertaken for the 88mm Pak 43/3 (Jagdpanther) and 88mm KwK 43 (Tiger B). Series maturation rested in the hands of the Krupp firm.

Recoilless (fixed) installation of the main gun

Stimulated by the Heereswaffenamt and Reichsministerium für Rüstung und Kriegsproduktion, Krupp and Rheinmetall studied the matter of a fixed installation of gun barrels in tanks and solidified their theoretical studies through practical testing.

In the time period from 28 to 30 April 1943 tests were conducted at the Unterlüss Firing Range using a Panzerkampfwagen II with a fi were available for testing on short notice. A 100 mm front plate and 60 mm side plates were welded to the existing armor plated walls of the vehicle. The baseplate socket-formed mounting for the ball end of the mortar barrel was welded on the front plate. Forward support of the barrel was arranged in such a manner that three elevations (0, 17 and 42 degrees) and three lateral points (0 and +/-15 degrees) were possible. Markings were placed on the vehicle, primarily on the suspension, to monitor the vehicle by slow motion photography during firing. Rolling resistance was set at 500 kg, total weight of the vehicle being 11 100 kg. Firing was initially conducted with a non-brakd running gear on flat concrete; later this was done with braked running gear and on grassy terrain. Preliminary firing results showed only minor movement of the vehicle during firing. The greatest movement was encountered when firing with non-braked running gear and amounted to approximately a 190 mm total recoil motion. The travel stroke was low, roughly 20 to 25 mm from a possible 130 mm. Aside from the recoil movement there were also vertical and lateral deviations, which were barely perceptible to the naked eye. One thing was certain, however, and that was that the motion of the Panzer II chassis was less than that of the self-propelled mount on the same chassis supporting the 75mm Pak 40 with recoil. The acceleration which occurred could easily be borne by crew members. It was shown that it was possible to mount a fixed gun with acceptable results and the vehicle was able to withstand great gas forces. During testing these forces averaged out to 130 t.

In order to provide a basis for solid development of such a piece, a barrel from a 75mm StuK 40 was installed fixed in the turret of a Panzerkampfwagen IV; this gun had a slightly higher impulse than that of the 120mm GraW 42 mortar. Additional trials continued at Kummersdorf on 18 November 1943. A 75mm L/48 barrel was mounted fixed in a Panzer IV. Gun mounts were strengthened. The 100 mm thick front plate was reinforced by a rib running its entire length. Even the elevating and traverse mechanism supports were significantly strengthened viz-a-viz previous testing. 80 rounds were fired without a muzzle brake. With the reinforced glacis plate, no problems were noted with the gun mounting and elevating and traverse mechanisms. The sighting device was inadequately fastened in place and broke from its mountings. A better solution was to be found by inserting springs or rubber cushions. The forces placed on the crew were quite tolerable. The vibrations which were carried through to the elevating and traversing wheels were much less than during previous firing trials, thanks to elastic mountings; however, these vibrations could not be tolerated over a period of time. With round 78 it was noted that a crack had developed in the gearbox housing of the Panzer IV and testing was therefore broken off. The design of a fixed, recoilless gun in a ball mount was thoroughly discussed at a meeting of the Panzerkommission on 21 December 1943.

On 11 January 1944 firing trials were resumed. A thorough study of a 75mm L/48 barrel mounted on a lighter Panzer II was to be conducted. In addition, a temporary mount was made for the gunsight (Sfl.Z.F.1a), the upper part of which had been broken off during previous firing; this piece was now supported by a clamping collar. 67 rounds were fired. During rapid firing the vehicle bucked somewhat from a weakening of the rear springs. The trials were encouraging and led to the development and testing of a tank destroyer with fixed barrel mounting on the Panzerkampfwagen 38(t) neue Art.

On 7 February 1944 the installation of a 88mm L/71 gun in a fixed position on a Panzer IV was completed at Krupp in Essen. The following advantages were noted: fighting compartment was roomier, more space for holding ammunition, aperture was smaller, manufacturing was simpler; a breechblock slider was necessary, however, with the fixed barrel.

Firing trials in Kummersdorf using a Sturmgeschütz with a 75mm L/48 gun without recoil were conducted on 7 March 1944. The gun was held in compass gimbals inside a 100 mm thick sloping front plate. This was reinforced by a vertical plate approximately 150 mm high and 20 mm thick on the outer front plate.

97 rounds were fired, during which the driver's periscope broke. The engine radiator sprung a leak. After each shot the Sturmgeschütz kicked back a few centimeters. The gun supports gave a very simple yet sturdy impression.

On 11 July 1944 testing was completed with a mass firing of 1000 shells. Various types of brackets for the sighting device were explored. Further improvements

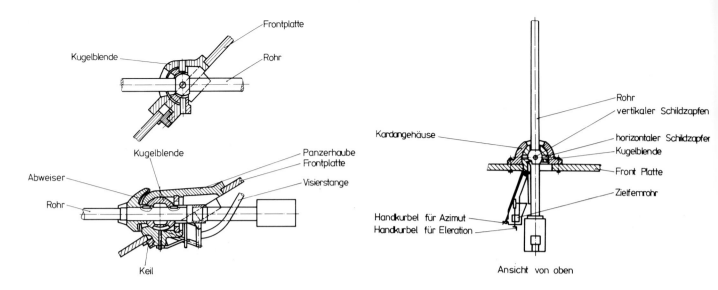

Drawings of the recoilless gun installed show the arrangement concept, which received much emphasis as the war drew to a close.

proved to be necessary.

By August 1st 1944 the Altmärkische Kettenwerk GmbH (Alkett) had built two prototype guns without recoil, which were installed in a Panzerjäger 38 and Panzer IV together with 1000 rounds. Even though difficulties cropped up initially, the last 600 shells were fired from the Panzer IV without further problems. With the Panzerjäger 38, however, there were more objections. Alkett wanted to introduce an entirely new design based on this experience. There was, however, a directive ordering the production of 100 Panzerjägers despite the previously noted difficulties.

On 11 August 1944 Krupp indicated its intentions of installing a fixed 75mm L/48 in the Panzerjäger 38, the 75mm L/70 in the Panzerjäger IV, and the 88mm L/71 in the Jagdpanther.

The contract for installation of the fixed 88mm L/71 gun in the Jagdpanther was awarded on 2 October 1944 to the firm of Krupp by the Heereswaffenamt/WaPrüf 4. Krupp proposed a Cardanic suspension for the gun, meaning the front plate of the tank was to be moved forward. The barrel was suspended in long shackles, which offered a certain amount of elastic shock absorption. The counterbalance was arranged in such a manner that it prevented any play in the gun mount in the direction of recoil. Advantages of this design were: space conservation, but only when a semi-automatic loader was not utilized; less material need; simpler manufacturing; smaller cut in the front plate. On 4 October 1944 Krupp design Hln-125 for the fixed installation of the 88mm L/71 gun was presented to the Heereswaffenamt/WaPrüf 4. The overall design was approved and cleared for actual construction.

The contract specified two prototype guns for installation in a Jagdpanther. Deadline for the first gun was the end of January 1945. The Heereswaffenamt/WaPrüf 4 provided guns with breech and elevating and traversing mechanisms; Krupp was left to produce the remaining parts as well as the assembly of the gun. The installation was to be conducted in a Jagdpanther of the MIAG firm.

For installation in the Jagdpanther, only the opening in the front plate would have to be changed accordingly. The recoil force was transmitted from the base plate via tie rods to a Cardan frame, which was held in place by vertical trunnion pins in its cast steel armor body. Particular attention would have to be paid to ensuring that this frame did not slide upward. Cartridge casings were ejected by a blowback-operated breech opener. In order to keep the acceleration forces from acting on the gunsight (W.Z.F. 1/2), it was fixed to the left side wall along with the elevating and traversing mechanism housing. Vertical and lateral control was accomplished by two parallelogram-type linkage units.

The unwanted deck openings were avoided by making use of this stationary sighting arrangement. Spindle elevating and traversing mechanisms of the simplest kind, belt-driven, provided an elevation and traverse range as on a recoil-type gun. Contracts were projected to be assigned as follows:
— Barrel, baseplate and — Dortmund-Hoerder breech, the first minus, Hüttenverein the second with blowback breech opener.
— Elevating and traversing — Hallesche mechanisms, counterbalance Maschinenfabrik
— Remaining components — Krupp
— Jagdpanther modification — MIAG
— Gun assembly — Krupp
— Installation in — MIAG Jagdpanther

Target date for completion of the development was expected by the beginning of January 1945.

The Heereswaffenamt/WaPrüf 4/II announced on 8 November 1944 that the contract number for the 88mm Pak 43/3 (two prototype models) would be SS 4902-0004-2096/44 (Gerät Nr. 5-08412, Contract 2961/44g). On 25 November 1944 Krupp informed the Mühlenbau- und Industrie AG (MIAG), Braunschweig, that a recoilless version of the 88mm L/71 gun had been shipped for the purposes of installation in a Jagdpanther. As it turned out, the cutout for the gun retaining plate required only minor changes for installing this new weapon. In addition, it was necessary to bolt the retaining plate on both its sides. The center-of-mass was shifted further back into the fighting compartment. When firing off a single round, the total acceleration of the vehicle amounted to 4.55 g; the total recoil of the vehicle was 65 mm.

On 14 February 1945 the Chef des Technischen Amtes (Chief, Technical Office) at the Reichsministerium für Rüstung und Kriegsproduktion, Hauptdienstleiter Saur, called for the immediate introduction of recoilless guns for the following armored vehicles:
— 75mm L/48 in the Panzerjäger 39;
— 75mm L/70 in the Panzerjäger 38;
— 105mm Sturmhaubitze in the Panzerjäger 38;
— 88mm L/71 in the Jagdpanther.

He called for the establishment of a study group, with Professor Dr.Ing. e.h. Wanninger of Rheinmetall as its chairman. All experience and results which had been acquired up to that point were to be compiled, evaluated and applied to further development.

On 12 February 1945 the arrangements for manufacturing the cast armor retaining plates for installation of the Pak 43/3 without barrel recoil in the Jagdpanther were begun. The cast steel parts were manufactured by Stahlwerk Mark in Wengern. According to the schedule, the first cast steel piece was to enter production on 20 March and be completed and ready for installation on 20 April 1945. From 5 March 1945 MIAG was in possession of a vehicle awaiting installation of the wooden mockup of the gun. Since the vehicle was part of the March production batch, the studies would have to be concluded no later than 15 March 1945. Due to the progress of the war, nothing further was accomplished regarding the installation of the recoilless 88mm Pak 43/3 in the Jagdpanther.

Bergepanther

Immediately after establishing the first Tiger and Panther battalions it was discovered that there was a dearth of suitable tow vehicles as a result of an oversight in planning. The reversal in the military situation, exemplified by the retreats beginning in 1943, made it increasingly necessary to recover damaged vehicles to be repaired by the maintenance units situated in the rear area. The shortage of adequate recovery means for Tigers and Panthers caused many of these vehicles to be left behind, either to be blown up by their own crew or to fall into the hands of the enemy. The Zugkraftwagen 18 ton half-track prime mover, which had proven adequate for recovering the Panzer III and IV, could only be used for towing the Tiger or Panther in tandem pairs or threes. When used in multiples, the frame of the Zgkw 18 t was seriously bent in spite of good cross bracing. The Maschinenfabrik Augsburg-Nürnberg was tasked with providing ten Bergepanther recovery vehicles minus winches and with a makeshift superstructure by no later than 6 June 1943. This demand by the Heereswaffenamt was only officially relayed to MAN on 7 May 1943. At the same time Henschel was assigned the task of producing the last 70 Panzerkampfwagen Panther Bergepanthers. Since the simplification of this vehicle during construction basically favored only the companies supplying the turret, Henschel saw this as an additional burden. The director of this factory, Dr.-Ing. Stieler von Heydekampf (also chairman of the Panzerkommission) declined the construction of the Bergepanther, but was nevertheless

The Pionier-Sonderfahrzeug (specialized engineering vehicle) developed by Maschinenfabrik Augsburg-Nürnberg was designed as an amphibious vehicle for creating obstacles and was to tow a heavy trench-forming plow.

ordered to manufacture this vehicle by the director of the Pz III subcommittee Dr.-Ing Blaicher.

The most pressing task in the design of the Bergepanther was the construction of a cable winch which would have a pulling capacity of 40 tons over a distance of 150 meters. Winches with this capability did not exist in the normal crane industry. It was, however, possible to make use of a winch which had been installed by the Maschinenfabrik Augsburg-Nürnberg back in 1940 in an amphibious Spezial-Pionierfahrzeug (specialized engineering vehicle). The military requirements for this vehicle called for an ability to destroy roads along their length and to set up

Bergepanther (Sd.Kfz.179) (initial design minus winch and spade)

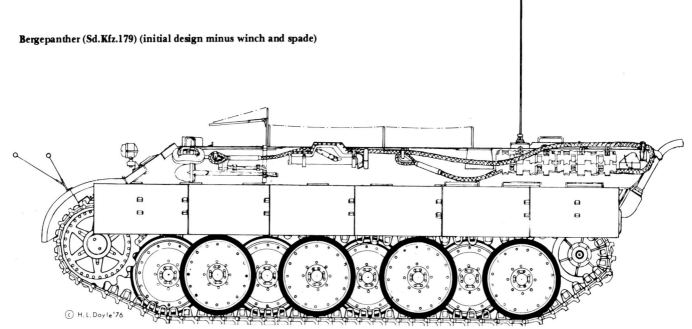

A 40 ton cable winch was mounted in the center of the vehicle.

Four of these Pionier-Sonderfahrzeuge were built based on the chassis of the 8-wheeled model of the Einheitsdiesel truck.

With minimal changes, the winch of the Pionier-Sonderfahrzeug was for the Bergepanther.

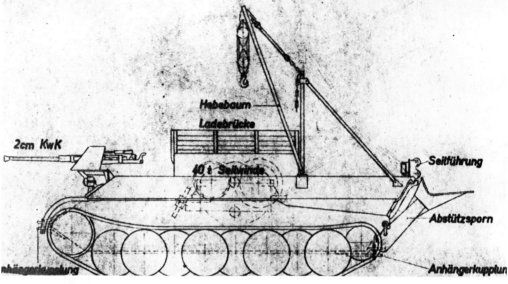

Rough sketch of the Bergepanther

Rear view of the Bergepanther with crane extended.

The spade in a nearly complete lowered position.

Side view of the Bergepanther with raised spade

Front view of the Bergepanther showing the shield for the 20mm gun.

Crew compartment of the Bergepanther. The 20mm gun bracket can be seen in the center front. The two flatiron supports on the front sides of the hull roof were for mounting an AA machine gun.

and remove obstacles. Two test prototypes of this vehicle, built up from the 8-wheeled model of the Einheitsdiesel truck, were completed in the fall of 1941. The installed winch could reel in an entrenching plow at a speed of 10 meters per minute. An earth spade was mounted at the rear of the vehicle in order to fully take advantage of the pulling force of the winch. Testing was carried out with satisfactory results; however, series production of this Spezial-Pionierfahrzeug did not occur.

The principle of this winch developed by MAN was also utilized for the Bergepanther, the winch being somewhat modified for installation in the Panther hull. The winch was of a vastly different concept than of the winches common to average hoisting devices. The winch, with the aid of eight drive pulleys mounted on two shafts, powered the tow cables by friction and wound the cable onto a drum behind the drive pulleys. The winch operated off of the power generated by the vehicle's engine via a gearbox

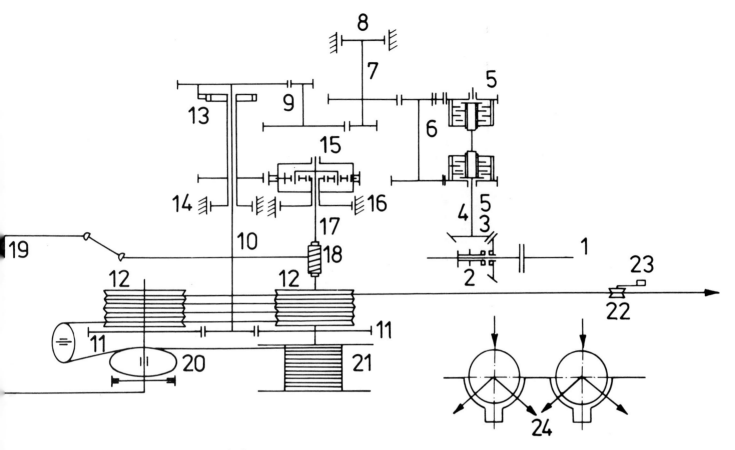

Schematic diagram of the Bergepanther cable winch.

1. drive shaft from vehicle engine, 2. mesh coupler (coupling and decoupling of the winch was only possible with engine not running), 3. bevel drive, 4. shaft for the clutches, 5. two multi-plate clutches for extending and retrieving the tow cable, 6. countershaft with tooth gears for right and left running, 7. jackshaft with main brake, 8. main drive brake, 9. main drive countershaft, 10. drive pulley shaft, 11. drive pulley gears, 12. drive pulleys, 13. ratchet drive (one-way) with bypass for cable winch drum, 14. cable winch drum brake, 15. rotary drive for cable winch drum, 16. power limiter for the cable drum (function of a slip clutch), 17. drive shaft for the cable drum, 18. worm gear driving the cable return, 19. crank gear operating the cable return mechanism, 20. cable return mechanism, 21. cable winch drum, 22. cable tensioning mechanism, 23. indicator light for cable tensioning mechanism, 24. synchronization of the tow cables into the drive pulleys.

and a connecting shaft. The construction of this high-strength winch was plagued with difficulties, since manufacturers in possession of the necessary machinery and knowledge were already involved in armament production and were already working beyond their capacity. It was therefore not possible to equip all Bergepanthers with winches. The winch compartment was externally encompassed by a plank-like structure and covered by a tent at the top. A heavy support spade was affixed to the rear of the Bergepanther, which was raised and lowered by the cable from the winch. During recovery, the tow cable was hooked onto the damaged vehicle; the Bergepanther drove out the entire length of the cable and then reeled the cable in using the winch, after the support spade had been anchored in the ground. Once the cable — and damaged vehicle — had been drawn in to the Bergepanther, the recovery vehicle then drove on after raising the spade above the ground. After the cable was let out again, the

40 ton cable winch as installed in the Bergepanther.

Cable winch removed from a Bergepanther.

Cable guide at rear of vehicle.

Details of the Bergepanther cable winch during testing with the British army.

Different designs of the cable guide were used.

These two pictures show the earth spade arrangement.

process was repeated until the damaged vehicle was clear of the greatest danger.

At the rear upper portion of the hull end a cable guide was mounted, consisting of two horizontal and two vertical rollers. This device made reeling the tow cable out and in much easier. A traction dynamometer was carried in the vehicle so as to prevent overloading the cable. Technical testing of the Bergepanther were carried out at the Kraftfahrversuchsstelle in Kummersdorf from 1944 on (under the direction of Major Dipl.-Ing. Wüst). The tow winch, after a few initial teething troubles, worked satisfactorily; the overall design was a success.

There were objections to the absence of a crane for replacing component parts on a damaged vehicle. The Verskraft in Kummersdorf accordingly developed a 2 ton auxiliary crane, which could be bolted onto all armored recovery vehicles as well as tanks. It was tested, but neither a general use nor a re-equipping of operational armored vehicles were possible due to the progress of the war. The auxiliary crane, a light jib, was bolted on at three points by threaded lugs either already set in the vehicle or fixed on later. With the Bergepanther or other vehicles having a fixed superstructure the crane swung through 120 degrees; on turreted vehicles the range was unlimited. At first, neither the towing coupling nor the tow rods of the Bergepanther prototypes were up to the rough troop operations in the field. Testing confirmed that simple linchpins were superior to all hinged and rotary type couplings. In the end, even the difficult requirement of ensuring the mobility of the towing apparatus in rough terrain was met.

To a great extent, the recovery of immobilized tanks depended on the availability of a nearby anchor point, the solidity of which would have to be greater than that of the tank being recovered. When using the cable winch for a simple pull, this function was basically fulfilled by the support spade. For recovery operations requiring multiple pulls the spade was generally inadequate. A ground anchor, transported as a two-wheeled trailer, was developed and tested for these difficult situations. During use the wheels were raised up. The effectiveness of the ground anchor as an anchor point was adequate, but as a trailer towed behind a tracked vehicle it was impractical; the design was not developed further.

Despite their good gripping strength, the tracks on the Bergepanther could not always be prevented from slipping during recovery operations in rough terrain. A "cable claw" was developed and tested for these temporary situations, the use of which permitted one or both tracks to be anchored to an anchor point, enabling the Bergepanther to drive ahead by one track ground contact length. For a single track link there were claw designs which could be inserted, bolted on, or hung. The difficulty lay in the fact that in order to make this method practical for front line operations, it would be necessary to change the overall design of the track accordingly. Even the tested grousers for snow and loose sand or the ice spikes would have required a modification of the track design. All these accessories would not only have benefitted the Bergepanther, but also would have proven advantageous for tanks as well. However, they were never employed operationally.

The Bergepanther carried a solid wooden beam on the left side of its superstructure for pushing damaged vehicles along or to one side. For this purpose two square-shaped steel plates were vertically welded to the bow of the Bergepanther.

In anticipation that the Bergepanther's role would not remain limited to just recovery and towing operations, but would also serve in the capacity as an equipment and energy supplier for meeting the material needs of tanks, other types of specialized tools aside from the auxiliary crane were developed and tested. Among these were a tension rack for track repairs, a hoist for replacing road wheels and accessories for "re-mobilizing" tanks with shot-up running gear.

The following memorandum from 15 June 1943 was found in the documents of the Henschel firm: "The contract initiated by Dr. Blaicher from the Heereswaffenamt WuG 6 for 70 Bergepanthers cannot be met by Henschel for the following reasons: The proscribed numbers of recovery vehicles to be delivered in 1943 is as follows: June 9, July 11, August 13, September 14, October 16 and November 7. According to a memo referring to a meeting with MAN from 28 May 1943, chassis construction including winch is expected to be concluded by the end of June 1943. Since a majority of the as yet unknown parts must be delivered and set in motion through manufacturing, it is entirely impossible to meet the shipping deadline set by the Heereswaffenamt.

A Bergepanther underway in service with the British forces.

A modified spade arrangement.

Shackles and pulleys, which comprised the standard equipment of the Bergepanther.

View looking down into the winch compartment. On the vehicle's right is a wooden block used for pushing damaged vehicles.

British forces gauging the pulling power of a Bergepanther.

The steel stops at the front of the vehicle were used to hold the wooden blocks carried by the Bergepanther and used for pushing broken-down tanks.

Mount for the 20mm gun on the forward hull of the Bergepanther.

A Bergepanther towing a Panther tank.

An Ausf. G Bergepanther still used by the French.

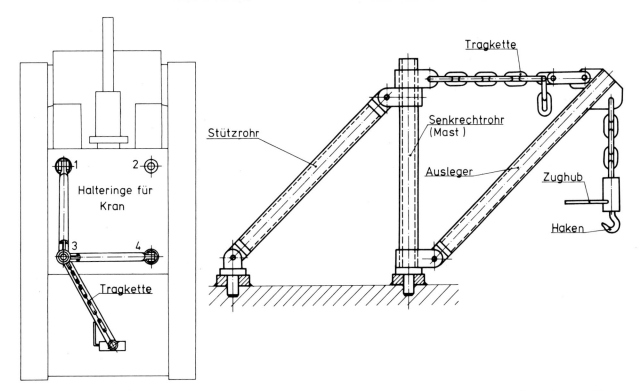

Auxiliary crane for armored vehicles, originally developed for assault gun units. The crane was detachable and when attached at points 1 and 2 was used to install and remove the transmission; at points 3 and 4 it was used for removing the engine.

If the 70 Bergepanther are to be tacked on to the end of Henschel production run of 200 Panthers, the first recovery vehicle would then be rolled out at the beginning of August 1943. Regarding this, Dr. Blaicher had clearly stressed in a meeting with Dr.-Ing. Stieler von Heydekampf (Henschel) that only the provisionary solution for the Bergepanther would be of concern for Henschel. The Heereswaffenamt/WaPrüf 5 IIIe, however, maintains that this is not a provisional arrangement, but the final solution in accordance with draft Tü 15 259. As a proposal by Henschel, it can only be reemphasized that without disrupting the Tiger II run we are only in a position to produce Panther chassis; these can then be converted to recovery vehicles at another site, such as at Krupp-Gruson in Magdeburg."

Nevertheless, the Bergepanther was produced by Henschel. On 10 July 1943 Henschel reported that in order to be able to deliver 11 Bergepanthers in July it would be impossible to supply five turret drives to Daimler-Benz, since these must also be installed between the engine and transmission of the Bergepanther; up to this point there had been no other type of element developed which could take over the job of transferring power.

The Bergepanther, Ausführung A (Sd.Kfz.179), made use of the Panther tank chassis with only minor modifications. The seats for the crew was at the front of the hull. A mount for accepting a 20mm Kampfwagenkanone for self-defense. The weapon was protected by an armor shield. To the right and left of the crew compartment

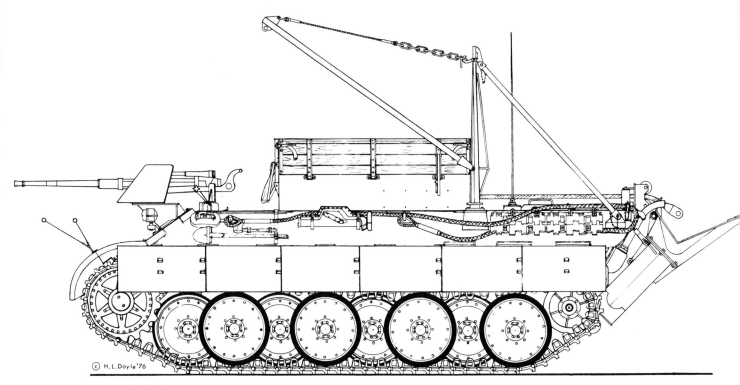

Bergepanther (Sd.Kfz.179) shown with planned equipment including 20mm KwK.

entry opening there were two brackets for affixing an anti-aircraft machine gun mount. The fuel supply was increased over that of the Panther tank to 1075 liters. The hull was manufactured by the firm of Ruhrstahl in Hattingen. Beginning in October 1943 the G model of the Bergepanther was shipped, which from March of 1944 was converted over to the hull model of the Panzerkampfwagen Panther, Ausführung G. In this production form the Bergepanther was produced up until the end of the war. Only after Demag, Werk Düsseldorf-Benrath, had taken over Bergepanther production was it possible to supply greater numbers of recovery vehicles (a total of 297, 46 of these without winch), albeit with great difficulties.

A Bergepanther with winch was demonstrated before the Generalinspekteur der Panzertruppen, Generaloberst Guderian at the Berka training range near Eisenach on 1 March 1944.

On 7 April 1944 Hitler was informed of the value of the new recovery vehicle by means of a series of photographs. It was announced that in addition to the measures already introduced for a monthly output of 20 Bergepanthers, in 1944 an additional 13 in April, 18 in May, 20 in June and 10 for July would be manufactured by converting Panther tanks from repair stocks. Manufacture of the 40 ton cable winch was accomplished at the firm of Raupach in Görlitz, later transferred to a new plant in Warnsdorf/Sudetenland. 20 to 30 winches were to be produced there per month. On 1 February 1945 an additional 125 Bergepanthers were projected for manufacturing. There was also a Gerät 549 in planning, a Bergepanther,

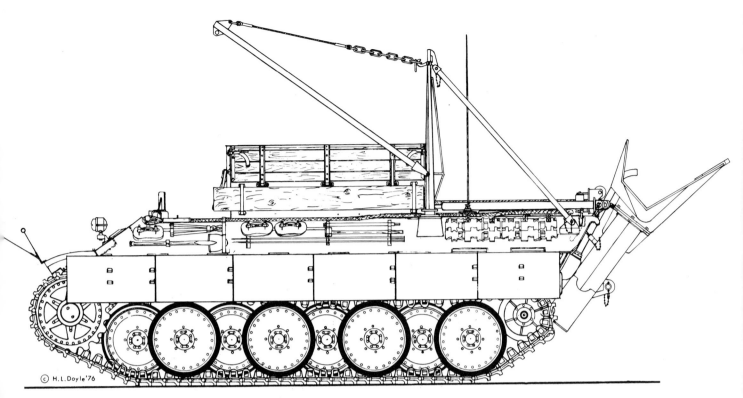

Bergepanther (Sd.Kfz.179) Ausf. G

Bergepanther (Sd.Kfz.179) final design

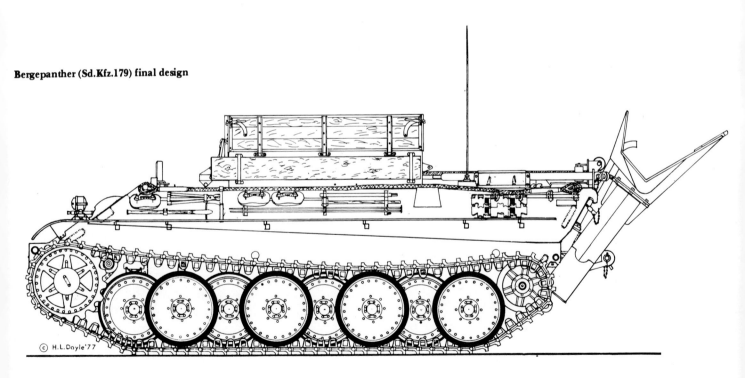

Ausführung F.

The Bergepanther provided valuable service in the recovery of heavy armored vehicles. Unfortunately, these were not available in sufficient quantity.

The front line troops also modified Panthers themselves by removing the turrets, using the vehicles for recovery purposes and as ammunition carriers on the battlefield.

It's also worth mentioning here studies undertaken for developing an armored mine-clearing vehicle and a Panther with a mine-clearing spade.

Flakpanzer

The development of the self-propelled armored anti-aircraft vehicle, or Flakpanzer, concept was under the direction of the Reichsluftfahrtministerium (Reich Aviation Ministry). As a result of draft SKA 593 Krupp carried out discussions as early as 2 September 1942 with engineers of the Generalluftzeugmeister regarding the redesign of the Flakversuchswagen (VFW) based on the guidelines of the Generalluftzeugmeister L Flak E4. Krupp proposed the use of the following tank components:
— Track and running gear of the Panther (Leopard armored reconnaissance vehicle running gear was not adequate, since the larger road wheels brought about an increase in the muzzle height by approximately 100 mm)
— Maybach HL 157 engine with 550 hp (Aufklärungspanzer Leopard)
— ZF AK 7-130 gearbox (Aufklärungspanzer Leopard)
— Steering mechanism from either the Leopard armored reconnaissance vehicle or, considering the great length, possibly an in-house design.

The combat weight of this Flakpanzer was to be roughly 31 tons, the muzzle height 2450 mm. A 1:1 scale wooden mockup was used to test all the platform arrangements and crew possibilities for the gun. The storage of on-board ammunition beneath the side seating as well as a reduction in size of the side platform plates were explored by Krupp.

On 4 November 1942 Krupp presented its drawings for a Flakpanzer (Gerät 42), making use of the following Panther components: running gear, track tensioning device, drive sprocket, final drive, steering mechanism, hand and foot levers, ZF AK 7-200 gearbox and Maybach HL 230 engine. Krupp relayed that the ZF SMG 90 gearbox and L 320 C steering mechanism, which were planned for use in the prototype model, had already been tested in another Krupp vehicle and that the gearbox exhibited serious problems, causing only limited driving performance to be reached. Krupp therefore proposed an alternate installation using the proven ZF SSG 76 gearbox (Panzer III and IV) together with its own design of a single-radius steering mechanism. Generalluftzeugmeister, L-Flak was in agreement with this recommendation.

On 20 January 1943, this department requested that the hinged side wall armor plating of the prototype's gun platform be attached in such a manner that they would offer partial protection for the crew against air attack.

The conversion of the self-propelled motor carriage for the Gerät 42 (VFW 2) using Panther components, based on draft SKA 716, was postponed until experience had been acquired with the prototype model. This approach presented no drawbacks, since no changes in construction design were foreseen compared with the prototype.

The situation in the air around mid 1943 is documented by committee report WK 2130 from 31 June 1943: ". . . enemy aircraft (tank busters or attack aircraft) presently attack using 30mm or 37mm cannon (later 50mm) at a range of approximately 1000 meters at 150-300 meters altitude. Dive bombers reach their lowest point in 1200-1500 meters altitude at an angle of 45-80 degrees.

Only directed defensive fire brings results. Vertical upward firing only serves as a scare tactic. In order to achieve a kill 60-70 rounds, at best 30 rounds, must be expected.

20mm is the limit, 37mm is better.

All defensive weapons must have a full 360-degree field-of-fire, given that at an airspeed of 500 km/h 140 meters can be covered in the space of one second. Armored vehicles protected on all sides can therefore not be utilized for flak roles.

When in their readiness positions, Panzer battalions of 96 vehicles cover an area 2 kilometers wide and 1.2 kilometers deep. Rheinmetall had proposed the following for the 30mm MK 108 (onboard anti-aircraft gun):
— attaching it to the tank's main gun, firing by using the barrel to an elevation of up to 20 degrees.
— mounting it on the rear part of the tank, possibly on the turret rear plate, giving a full range of elevation.
In the view of General von Renz, it was possible to provide a defense out to a range of 600 meters using a machine gun. Against aircraft, which release their multiple weapons from an altitude of 1000 to 1500 meters, even the 20mm gun is hardly sufficient.

Attack pilot Oberstleutnant Weiss reports that the anti-aircraft protection of the Russian tank division has become rather effective at warding off our aircraft, since the material losses are too great.

General von Axthelm recommends the use of supplemental Siemens-Martin steel on the 12 t Zgkw half-tracks for protection against armor-piercing rounds, since the flak guns these half-tracks tow are increasingly being called upon to provide anti-aircraft protection for the Panzer battalions when the aircraft are forced to higher altitudes by our medium flak guns.

Conclusion of the Waffen- und Panzerkommission: The installation of anti-aircraft weapons in the available tanks does not appear possible. The use of currently available machine guns in tanks for anti-aircraft self-defense when in readiness positions and on the march is planned. Initially, the use of the Panzer IV as a Flakpanzer will be studied. If this does not prove adequate, we will have to fall back on the Panther chassis. The following solutions are considered for the armament:
— on the Panzer IV:
a) 20mm quad
b) 37mm single
— on the Panther: a) 20mm quad (minus carriage)
b) 37mm twin or triple
c) After development has been concluded, possible use of 55mm twin gun.

signed, Nill..."

At a meeting on 8 June 1943 concerning the development of a Flakpanzer using an 88mm Flak 41 it was determined that the conversion work on the Panther chassis (available for conversion to a Flakpanzer Flak 41 in September 1943) would begin after design work had been concluded on the improvised gun.

During a conference at Krupp-Gruson in Magdeburg on 21-22 October 1943 regarding the self-propelled Flak 41 (VFW) the Generalluftzeugmeister GL/Flak requested a timely completion of the prototype so that it could be demonstrated on at the Wustrow anti-aircraft firing range near the Baltic Sea resort of Kühlungborn. The basis for the development of the VFW II was to be the Panther II chassis. The Panther I chassis, already delivered, was to be returned accordingly. The Reichsminister der Luftfahrt und Oberbefehlshaber der Luftwaffe, Generalluftzeugmeister, informed the Krupp firm on 21 December 1943 that the following contracts were valid for self-propelled light and heavy anti-aircraft guns:
— Contract GL/Flak — E4/V -DE — 0084 -6710/42
— Contract GL/Flak — E4/V -DE — 0084 -6715/42 (heavy gun)
— Contract GL/Flak — E4/V -DE — 0084 -6710/42 (light gun)

Additionally, the Generalluftzeugmeister confirmed that in place of the heavy Flakpanzer VFW II (design contract DE-0084 — 6715/42) only the light Flakpanzer VFWL (design contract 0084 — 6717/42) would be developed.

According to information from Krupp on 13 January 1944, Daimler-Benz, Berlin-Marienfelde, was developing a turret for a 37mm twin gun which was to be set on a Panther chassis. Frontal armor 100 mm, side armor 40 mm. The power for the turret traverse came from the vehicle's engine via a hydraulic drive. There were, however, considerable problems with the design. A prototype was to be ready by the middle of 1944. Rheinmetall also had to overcome serious problems with the construction of a turret for the self-propelled 88mm Flak 41.

On 13 January the Reichsministerium für Rüstung und Kriegsproduktion issued the following directive: "The firms of Krupp, Essen, and Rheinmetall, Düsseldorf, are to cease development of the self-propelled 88mm Flak 41 for the Panzer IV and Panther I, as there is more urgent developmental work to be done." Krupp-Gruson reported on the state of the Flakpanzer development on 27 January 1944. The final decision was a design of Krupp, and involved setting a 37mm twin gun on a Panther chassis in place of the previously proposed Daimler-Benz turret.

There were serious doubts raised as to whether this Flakpanzer would even prove suitable for use. It was difficult to draw off the powder gasses, to control secondary ignition of the cartridge ammunition and to properly lock on to the target. A remedy presented itself on the horizon, however, as there were already plans to install the 30mm triple-barreled gun if the 30mm twin gun was not available by the fall of 1944. In any case, a test model was not to be expected before the end of 1944.

Around the end of January 1944 feasibility studies were undertaken to replace the 37mm twin gun with a 55mm gun. A mockup turret was built for Gerät 554 (Flakzwilling 37mm) and set on a Panther chassis. Preliminary work was carried out by Rheinmetall-Borsig under the cover-name "COELIAN." The installation was planned as a 37mm Flakzwilling 341 twin-barreled gun with a rate-of-fire of 2x500 rounds per minute and a muzzle velocity of 1000 meters per second. The weapons were installed beneath armor protection and had a full 360 degree traverse.

A mockup turret for the 37mm Flakzwilling 341 anti-aircraft gun fixed to a Panther D chassis. Designated "Coelian"

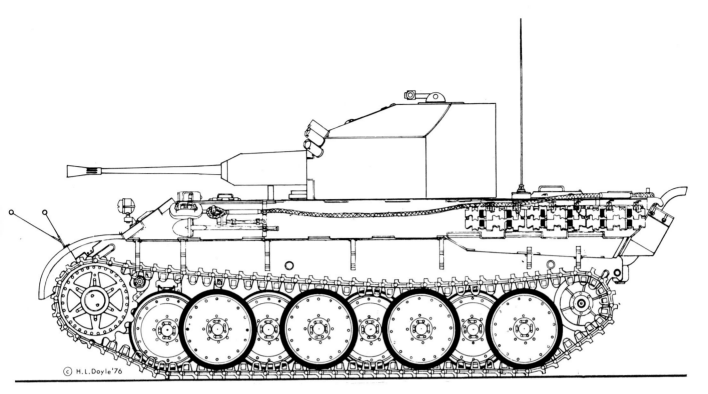

Flakpanzer Panther "Coelian" with 37mm Flakzwilling 341

Flakpanzer Panther (with barrels at greatest elevation)

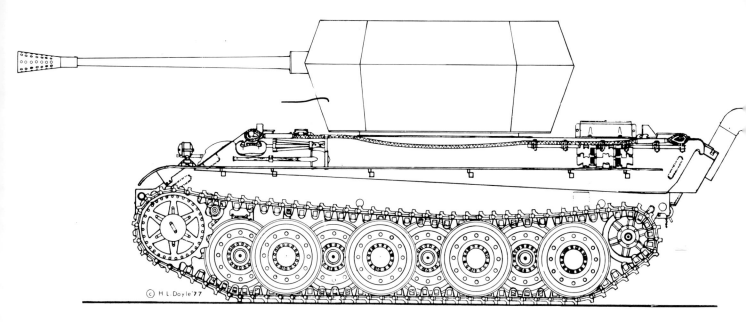

Flakpanzer Panther with 55mm Flakzwilling Gerät 58

On May 8th 1944 it was announced that the General der Flakwaffe required a Flakpanzer with a 55mm anti-aircraft gun. Since a new chassis design was out of the question, the Panther chassis would be utilized — if possible without changes. The use of hinged platforms, such as found on the 37mm Flakpanzer, was not considered. A traversable cupola or a turret-like design was preferred. Krupp was assigned the task of carrying out this developmental work.

Study results for the installation of a 37mm Flakzwilling 44/L 57 in an armored turret were presented during a meeting at the Heeresversuchsstelle für Panzer und Motorisierung in Kummersdorf-Schiessplatz between the Heereswaffenamt/WaPrüf 6 and the firm of Vereinigte Apparatebau AG on 16 January 1945. The range of the gun (discounting trajectory) was 6600 meters; the maximum altitude was 4800 meters. The Panzer IV turret bearing race was utilized in accordance with draft H-Sk A 90 360.

Upon presentation of the design the Vereinigte Apparatebau AG was informed that the Generalinspekteur der Panzertruppen had rejected the installation of the 37mm weapon, since the firepower-to-vehicle weight ratio was too low. With this, the studies came to a conclusion. Nevertheless, this project was pursued further and attempts were made to accommodate the Flakzwilling 44 on the chassis of a Panzerkampfwagen IV or the Panzerjäger 38 (D).

On 17 February 1945 a wooden mockup of the 55mm Flakpanzer, Gerät 58, was sent from the Seidenberg branch of the Vereinigte Apparatebau AG to Grimma in Saxony. There it was reassembled and shown to the Heereswaffenamt. The Heereswaffenamt/WaPrüf 6 laid down the requirement that the necessary pumps driving the pressurized hydraulic motor for the elevating mechanism were to be arranged in a series inside the fighting compartment. In addition, they were to be optionally driven by the main engine in order to ensure the operational readiness of the Flakpanzer at all times.

An auxiliary power supply was originally intended to be a motor made by the Hirth company, but no additional developmental work had been done up to that time. A Volkwagen engine was therefore suggested, the measurements of which, however, were too large to fit into the fighting compartment. In addition, the feed for the cooling air proved inadequate. A continuous motor output of 21 hp was necessary for driving the two pressurized hydraulic pumps; 42 hp for 3 to 4 seconds was needed during acceleration when the vehicle was at an angle of 10 degrees. Since the VW engine only had a peak performance of 32 hp, it was insufficient for acceleration. Moreover, by installing the engine, space was lost for storing ammunition. The danger of the Maybach primary engine having poor starting qualities at extremely low temperatures was alleviated by the use of a pre-heating apparatus. The idea of an auxiliary power unit was therefore dispensed with. Since the two pumps for the turret traverse drive could not be installed in the engine compartment (due to space constraints), these were carried —together with the oil reservoirs — in the turret basket. Power for the disengageable pumps was accomplished by means of a gearbox off the cardan shaft to a hydraulic drive of the turret traverse mechanism, as on the Panther. The housing for the turret drive remained unchanged.

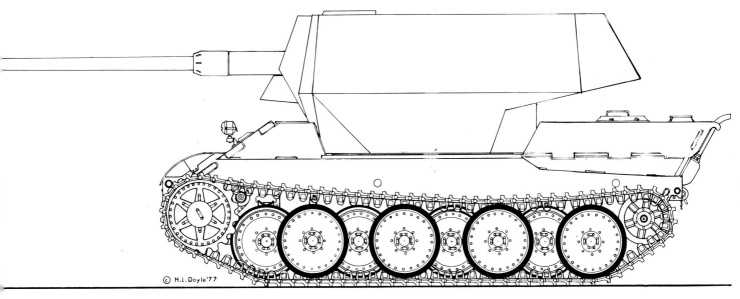

Flakpanzer Panther with 88mm Flak 41

Wooden model of a mobile anti-aircraft gun on a Panther chassis using an 88mm Flak 41

Since the developmental work on the 55mm Flakpanzer was not included in the emergency program, it was dropped. Preliminary work was brought to a particular state of affairs and concluded.

There were also plans to mount an 88mm Flak 41 (range 20,000 meters, maximum height 14,700 meters), with a 360 degree traverse range, on the chassis of the Panther. The gun was protected in a housing enclosed on its sides by armor plating and open on top.

Of this Flakpanzer there only remains photographs of a wooden model.

As opposed to the Krupp-built VFW (Versuchsflakwagen, or test anti-aircraft vehicle), the following comparative data is provided for the planned series vehicle:

	Test models	Series
weight with crew	25000 kg	31000 kg
engine	Maybach HL 90	= HL 157
engine output	360 hp	550 hp
engine r.p.m.	3600 1/min	3500 1/min
specific engine output	14.4 hp/ton	17.5 hp/ton
gearbox	ZF SMG 90	= AK 7-130
steering unit	Henschel L 320 C	MIAG/Leopard
running gear, arrangement	BW interleaved	Panther
roadwheel diameter	700 mm	860 mm
individual wheel load	1560 kg	1940 kg
armor:		
body, front/side	20/14.5 mm	30/16 mm
protective shielding, front/side	12 (below 45 degrees)/8	
maximum speed	60 km/h	60 km/h
total length minus gun	6700 mm	6700 mm
total length with gun	9830 mm	9900 mm
total width without/with platform	3000/6000 mm	3270/6520 mm
total height	2800 mm	2920 mm
muzzle height	2220 mm	2450 mm
hull width, interior	1800/2080 mm	1800 mm
track ground contact length	4000 mm	3920 mm
track width	420 mm	660 mm
track base	2580 mm	2610 mm
ground clearance	400 mm	540 mm
specific ground pressure	0.6 kg/cm²	0.58 kg/cm²
climb	28 degrees	30 degrees
trench	2500 mm	2500 mm
vertical step	700 mm	820 mm
wading depth	1100 mm	1300 mm
fuel supply	600 liters	550 liters
number of fuel tanks	4	4
range on paved road	300-350 km	300-350 km
range over terrain	200-250 km	200-250 km
number of rounds permanently carried	36	36
number of rounds in satchels, approximate	9	10
operating space, diameter	5000 mm	5000 mm
hull weight	6000 kg	6500 kg
elevation range	-5 to +90 degrees	

lateral range:
1. with closed platform = 40 degrees
2. at platform opened to 30 degrees and with 10 degree elevation = 360 degrees
3. with platform completely opened at -3 degrees depression = 360 degrees

adjusting rate for elevation	1. Flak 2-4 degrees per hand crank
	2. Pak 0.2 degrees per hand crank
adjusting rate for traverse	1. Flak 100/6400
	2. Pak 50/6400
recoil	1000 mm
buffer force	at 0 degrees elevation = 5800 kg
	at 90 degrees elevation = 7700 kg

Self-propelled artillery

Planning for self-propelled artillery began in the spring of 1942 when the Heereswaffenamt awarded contracts to Krupp and Rheinmetall for the development of these types of vehicles. Krupp recommended on 4 June 1942 that as many components as possible from the Panther be utilized for the Panzerhaubitze (self-propelled howitzer) project. Krupp asked the Heereswaffenamt/WaPrüf 6 for comprehensive drawings and a weight breakdown of the individual Panther components. At a meeting on 4 September 1942 Krupp proposed that, contrary to previous thinking, the Panther chassis should be used for the leFH 43 Panzerhaubitze as well, since unlike the chassis of the Aufklärungspanzer Leopard it had a more favorable design. A weight increase of 11 metric tons was to be expected. Using a wooden mockup of the Panzerhaubitze it was decided that the gun would have: a breech opening to the left, semi-automatic and electrically fired. After an examination of the space inside the fighting compartment, it was decided to dispense with the current arrangement of ammunition storage (cartridge rounds) in order to accommodate a second loader. The question of where the ammunition would finally be carried would be left until the finished wooden model. Krupp projected completion of this model for the beginning of October 1942. There was already the need to install a machine gun in the glacis plate of the 1:1 model. In a makeshift arrangement, Krupp also attempted to place a leFH 43 and the 88mm Pak on the Leopard chassis in the Grille Panzerhaubitze shape without making any major design changes.

According to draft W 1734, the Panzerhaubitze sFH 43 (self-propelled) using Panther components had a combat weight of 34 tons and a muzzle height of 2700 mm. The muzzle height was primarily dictated by the engine. The planned barrel depression of 8 degrees through a complete revolution of the turret was to be maintained; studies had shown that limiting the gun to a 3 degree depression would only reduce the muzzle height by approximately 130 mm. A loading apparatus was planned for calibers 12.8 and 150mm. A feed rammer was considered for the shells, which was controlled in a telescoping fashion. The project made use of the new design of the 150mm howitzer, which on a wheeled carriage had a muzzle height of 1475 mm in a spread trail position and 1220 mm in a fixed panoramic position.

After viewing the wooden model of the Panzerhaubitze sFH 43 (Grille 15) and the 128mm Panzerkanone K 43 (Grille 12), the Heereswaffenamt/WaPrüf 6/III proposed

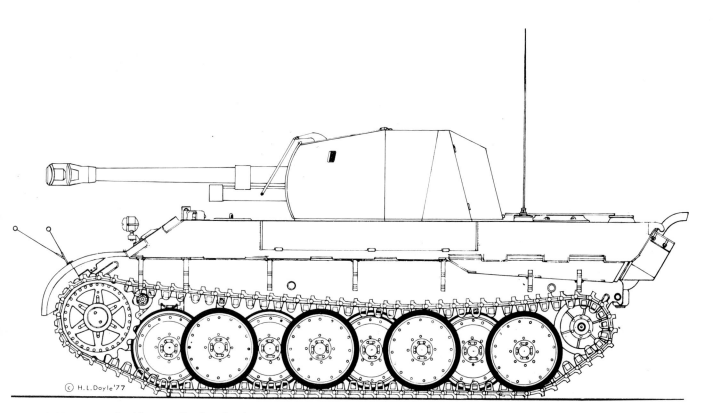

105mm leFH 43 Panzerhaubitze on a Panther chassis

lowering the muzzle height from 2670 to 2500 mm by:
— lowering the engine compartment through changes in the cooling system and shifting the combustion air filter. Later it was determined that it was not possible to drop the Maybach HL 230 engine, only the Maybach HL 210;
— Installation of the shorter Tiger OLVAR in place of the ZF AK 7-200 gearbox;
— Shifting the center point of the gun roughly 300 mm by rotating the pivot plate 180 degrees.

Since the apparatus for separating the gun from the chassis could not be accommodated in the vehicle, this would have to be carried along in a transporter. Not every Panzerhaubitze was given a separating device. Efforts were to be made to mount the three outriggers, each weighing 400 kg. If these were carried on the track sponsons the angled sides of the superstructure could be eliminated, since the outriggers offered additional shot protection. The Heereswaffenamt/WaPrüf 4 required that the gun could be rotated and fired through 360 degrees with the hinged platform sides in the closed position. An additional requirement was the ability to use the driving brakes when firing so as to prevent the Panzerhaubitze from rolling back when firing from a hull-down position. Approximately ten rounds of ammunition were to be carried on the platform, with an additional 20 stored in the rear of the vehicle. Ammunition could also be carried on the platform when travelling.

Between the Heereswaffenamt and Krupp the following guidelines were established: the gun crew consisted of four men. Tank seats were to be used as seating. For sighting, a tank gunsight was to be used in addition to a rotating periscope. WaPrüf 4 recommended the use of Siemens-Martin steel (50 mm front, 20 mm sides and rear) instead of armor or electric furnace steel (30 mm front, 16 mm sides and rear). Krupp calculated that a 1.5 ton increase in weight would result from the use of Siemens-Martin steel. WaPrüf 4 was in agreement with the use of armor plating then available for the prototypes. Siemens-Martin steel would only be available beginning in 1944, and even then there was the possibility that 30 mm armor plating would be used for the front plate. On 3 December 1942 the Heereswaffenamt/WaPrüf 4/IIIb rejected the use of Panther components for the LeFH 43 Panzerhaubitze 43, meaning that the components would be derived from the original proposal, the Leopard. In conjunction with this decision, it was accepted that the deadline of 1 April 1943 for the completion of a prototype could not be met.

In order to conduct firing trials as soon as possible in spite of this obstacle, a prototype using components from a Panzer IV was to be built. Only those parts of the Panzer IV were required which would be used in the construction of the prototype. Continued development of the Panzerhaubitze sFH 43 (Grille 15) and the Panzerkanone 128mm (Grille 12) was to be emphasized. The Heereswaffenamt once again renewed its requirement that the gun be separable from the chassis. Krupp therefore proposed that

The light field howitzer was mounted on a Panther chassis and enjoyed a full 360 degree traverse.

this separation be accomplished by slide rails or a similar arrangement, since using a jib or crane would lead to problems, given the great weight involved. A separating apparatus integral to the self-propelled vehicle was indispensable. By 26 January 1943 the Heereswaffenamt had assigned the following contracts for the use of Panther components in developing Panzerhaubitzen and Panzerkanonen:

— WaPrüf 6/III SS 006-6570/42 for self-propelled artillery sFH 43 (Grille 15)
— WaPrüf 6/III SS 006-6572/42 for self-propelled artillery 128mm K.Kp. (Grille 12)
P223:
— WaPrüf 6/III SS 006-6568/42 for self-propelled artillery (Heuschrecke 12/13)

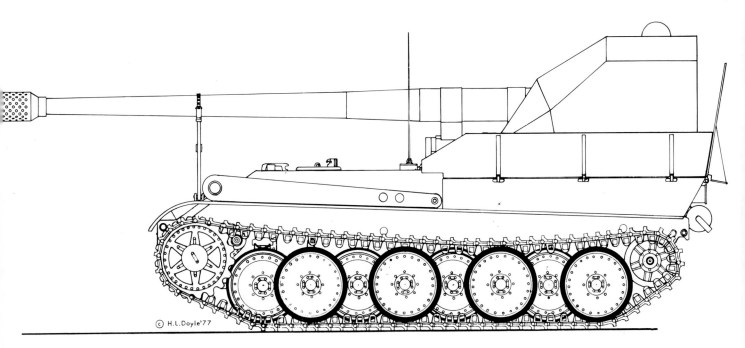

128mm Panzerkanone Gerät 5-1211 Krupp

Since difficulties (and therefore reduced output) were anticipated in converting the sFH 18 (150mm) and leFH 18 (105mm) over to the sFH 43 (150mm) and leFH 43 (105mm) respectively, on 11 March 1943 the WaPrüf 4 called for a study to determine the feasibility of installing the sFH 18 and the leFH 18 in the Panzerhaubitze Grille. The previous requirements (360 degree field of fire, upper elevation group, separable gun) would remain in effect.

In the meantime, since the armor for the Panzerhaubitzen would have to be made of Siemens-Martin steel, the projected enclosed armor was too heavy and the 360 degree traverse too problematic. The Heereswaffenamt/WaPrüf 4 therefore called for an open top armor with sides sloped to the rear. In this manner, it was hoped to achieve a refined shape to the superstructure. For construction of the prototypes, components of the Panther I

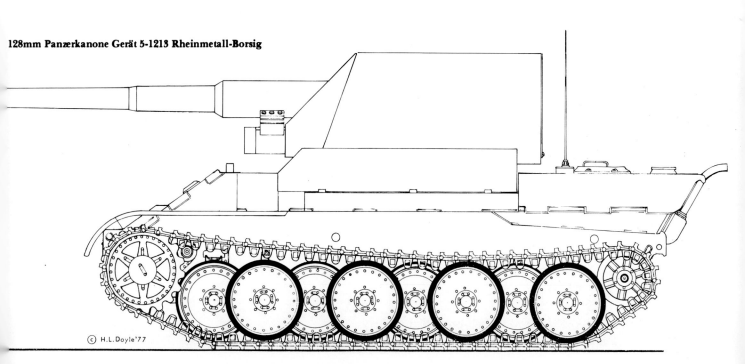

128mm Panzerkanone Gerät 5-1213 Rheinmetall-Borsig

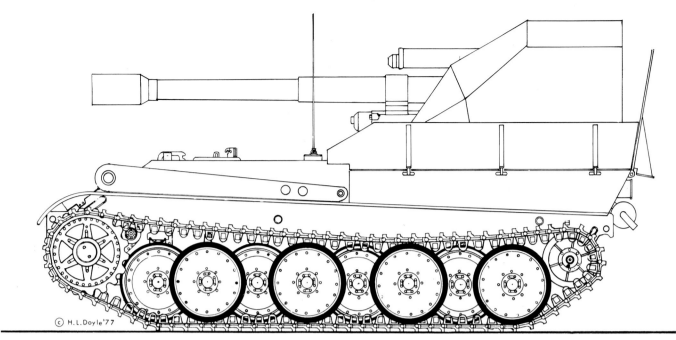

150mm Panzerhaubitze Gerät 5-1528 Krupp

would be used. For series production, however, the Panther II was to serve as the foundation. A standardization with the Flakpanzer Gerät 42 (88mm Flak gun) was to be a goal.

On April 20th 1943 it was discovered that the use of the sFH 18 150mm barrel for the Grille had run into difficulties. The decision was then made to continue the design of the 150mm sFH 43 and 128mm Sfl. in such a manner that these would incorporate as many parts of the sFH 18 as possible. Of primary consideration was the construction of the Heuschrecke 15, with its gun set to the rear, where separating the gun from the body would be accomplished from the idler wheel by means of rotating shafts. In order to gain practical experience for this remedy, preliminary tests were done at Krupp-Gruson in Magdeburg on the Sfl IVb which had been provided.

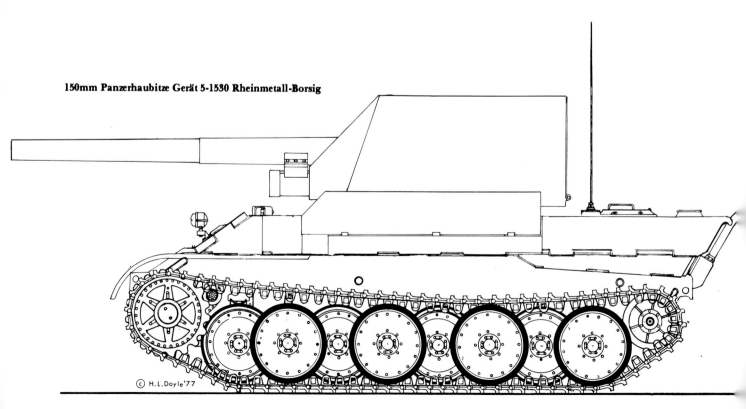

150mm Panzerhaubitze Gerät 5-1530 Rheinmetall-Borsig

Utilizing parts from a Panther chassis Krupp developed a self-propelled gun based on the 128mm Kanone 43. A similar idea was submitted for the 150mm heavy field howitzer as a self-propelled howitzer.

Sturmmörser 150mm assault mortar

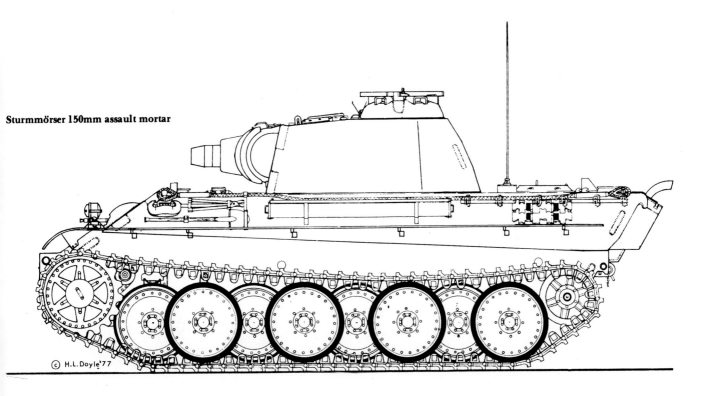

Heereswaffenamt/WaPrüf 4 indicated that development of the 128mm self-propelled gun was urgent. It appeared necessary to employ the same construction to the 128mm Kanone as was already being used with the Sturmkanone and Maus, as their projected jacketed cradle was more appropriate for the Heuschrecke project.

On 3 June 1943 MAN announced that additional Panther components for the 128mm K. 43 (Sfl) and sFH 43 (Sfl) would only be available starting in August of that year.

Initial project drawings for the Panzerhaubitze sFH 18/3 Geschützwagen V (Panther chassis) were presented on 11 February 1944:
— composition of the Panzerhaubitze in accordance with draft Bz 3234
— sFH 18/3 gun, separated, in accordance with draft Bz 3236
— separating procedure according to Bz 3237
This Panzerhaubitze was similar to the leFH 18/6(Sf)/Gw III/IV. The muzzle brake was dropped due to the requirement for firing cartridge case base (TS) rounds.

With the absence of the muzzle brake the recoil force increased to 28 metric tons, a force which, in the view of the Heereswaffenamt/WaPrüf 4, could be absorbed by the vehicle. The sFH 18/1(Sf)/Gw III/IV (Hummel) was used as a comparison; with a combat weight of 24 tons it was able to take the shock of a 22 metric ton recoil force.

The elevation when firing from the vehicle would have to be increased to +45 degrees, requiring an increase in the muzzle height from 2250 to 2400 mm. With the gun removed from off the chassis, the powerful recoil pressure prohibited the use of a firing framework. In its place was a type of round bottom carriage, which was lifted off the chassis together with the turret and set onto its four sets of outrigger pairs. Removal was accomplished in the following manner:

The gun was turned to 90 degrees. The left side plate of the Panzerhaubitze was folded down, forming two guide rails which ran perpendicular to the vehicle's direction of travel. At the end of each rail a spar was fixed running vertically upward. Both spars were given a crossbeam reinforcement at their upper ends. A roller block was fixed onto both the right and left sloping side walls of the turret. With the mounting of these roller blocks the turret was easily lifted by a hoisting action. It could then be smoothly moved on the blocks' four rollers. The barrel was elevated, the turret raised with the aid of two tackle blocks and the roller blocks were removed. Then the turret was lowered, and the four outriggers were assembled. The four outrigger pairs were carried on the vehicle in front of and behind the turret.

On 22 February 1944 the Heereswaffenamt/WaPrüf 4 required that the number of outrigger pairs be reduced from four to three. In this manner it was possible to keep the muzzle height of the separated gun lower. In addition, a three point arrangement seemed to offer greater advantages for the separated gun. There was an inquiry to determine how far it would be possible to mount the 128mm K.44 in the cradle of the sFH 18/3 (Sf). On 18 March 1944 a 210mm mortar was called for mounting on a Panther chassis. On 23 June 1944 in Kummersdorf Krupp handed over draft O U 23550 for the installation of the 150mm Sturmmörser (assault mortar) in the turret of a Panther. At the same time Alkett was working on a similar project — but proposed the use of the Sturmgeschütz chassis. Among other things, the Krupp design revealed a range-finder and gunsight. The commander's cupola was taken from the Tiger, the gunsight from the Maus. The range-finder, for a base of 2300 mm, would have to be a new design.

On 21 July 1944 Oberst Dipl.-Ing. Crohn, Hauptgruppenleiter in the Heereswaffenamt/WaPrüf 6, informed Professor Dr.-Ing. Müller of Krupp that due to the current military situation all development of armored self-propelled vehicles for heavy and super-heavy guns was to be stopped. Nevertheless, the wooden mockup of the Sturmmörser 150mm was to be demonstrated before the Chef des Generalstabes des Heeres, Generaloberst Guderian. Even the recommendation to set the 210mm mortar on the Panther chassis was halted; already 100 design hours, at RM 32.25 each, had been devoted to this project. On 23 August 1944 Oberst Crohn called for the continuation of the Sturm-Panther project. He was of the opinion that there were significant advantages to the panoramic field of fire. The Heereswaffenamt/WaPrüf 4 indicated on 22 September 1944 that even the development of the Panzerhaubitze sFH 18 on the Panther chassis was to be stopped, given that it was no longer cleared for its primary purpose.

Rheinmetall had planned to make use of the Panther chassis for similar developments for Panzerartillerie. The equipment designators were:
— Gerät 5-1213 (128mm K (Sfl) RhB);
— Gerät 5-1530 (sFH 43 (Sfl) RhB)

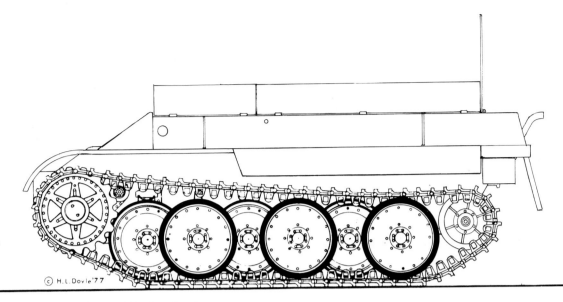

Ammunition carrier based on a shortened Panther chassis

Krupp designators were as follows:
— Gerät 5-1211 (128mm K 43 (Sfl) Kp I)
— Gerät 5-1528 (Sf Kp I)

In the official records there was also a Gerät 811 — a Geschützwagen (Gw) Panther for the schwere Feldhaubitze 18/4(Sf) based on AZ 735 WaPrüf 4/Is from 6 July 1944, and a Gerät 808 Geschützwagen V for the sFH 18/2(Sf) — which according to a directive from 19 November 1944 was dropped because the plans were not ready.

The developments of the Krupp firm were conducted under the collective designation "Grille", while those of Rheinmetall-Borsig ran under "Skorpion." The Panther self-propelled vehicles were divided into two major classes:
— Size 10 for the 88mm Flak, 105mm leFH or the 100mm Kanone.
— Size 15 for the 128mm Kanone or the 150mm sFH.

Together with Rheinmetall, Daimler-Benz developed an ammunition carrier which was to be built from the shortened chassis of a Panther.

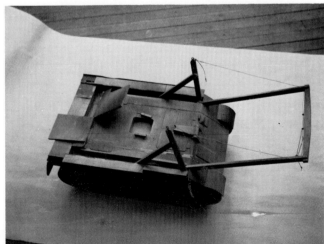

In 1943/44 a wooden model of an ammunition carrier was developed by Daimler-Benz, Berlin-Marienfelde, with Rheinmetall-Borsig, based on the Panther chassis and later built as a mild steel prototype. The Panther chassis was shortened by the length of one road wheel pair. The Heereswaffenamt/WaPrüf 4 called for this to be developed as an artillery tractor and self-propelled vehicle. The gun which it would carry could by hoisted on and off by a crane fixed to the vehicle and could also be fired from the ground. The chassis of the gun carrier could then be operated as an ammunition carrier or tow vehicle. The thickness of the armor plating ranged from 20 to 30 mm.

Near the end of the war Skoda developed a 105mm multiple rocket launcher. There were plans to accommodate this design on a Panther chassis, among others. The gun mounting of the 88mm Flak was to be used. The firing apparatus, made of angle iron welded together, was kept simple, the length of the firing chamber being 3500 mm.

A 105mm rocket weighed 19 kg. With a full 360 degree traverse range, the apparatus had an elevation of -5 to +75 degrees. When ready to fire, the launcher weighed roughly 3.5 tons.

Since it was no longer possible to sufficiently equip the front-line troops with Panther tanks and replacement parts, the development of Panzerartillerie on the Panther chassis did not continue beyond drawings, wooden models and a few prototypes.

105mm mobile multiple rocket launcher

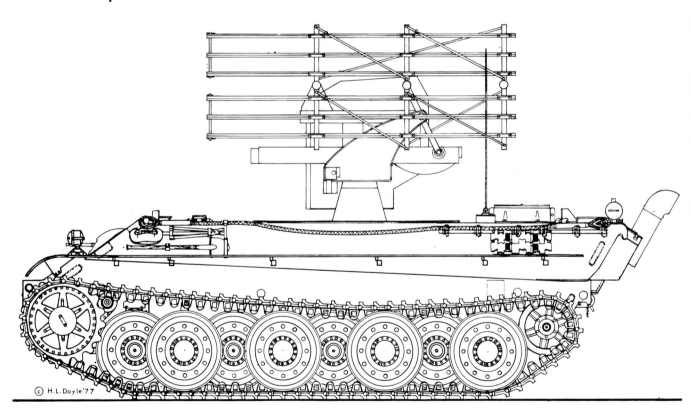

Conclusion

The Panzerkampfwagen Panther was, technically speaking, a unique accomplishment from the both the Army and armament offices responsible as well as the industry. Even if not all its weak points could be eliminated due to reasons of design manufacturing, in the last years of the war the front-line troops had mastered the Panther to the point where it could take on any enemy tank with an excellent prospect for victory. It possessed an outstanding gun firing a flat trajectory, the superior frontal armor and its mobility in terrain thanks to an acceptable performance weight, not to mention the superb running gear and its advanced suspension. As opposed to these, there was its above-average height, weak side armor, inadequate range as well as an unreliable engine and final drive. Coupled with these problems was an insufficient production of replacement parts for a vehicle which was forced into mass production before reaching technical maturity.

Nevertheless, the Panther compares favorably even with the postwar combat tanks still in use by today's forces.

Had it been possible to install the improvements being tested, such as (among other things) the fuel-injected gasoline or diesel engine, the electric, hydrostatic and hydrodynamic transmissions, gyro-stabilized gun and gunsight, automatic loader, infra-red night sighting and driving devices, this tank would have proven to be an opponent to be reckoned with even today. The fact that two Panthers could be built in the same amount of time it took to produce a single Tiger was a point that clearly shifted the focus of production to the Panther. A monthly output of 400 of these vehicles was projected.

Again and again it was shown that the continuing changes in the conception — calling for a Panther II even before the teething troubles had been alleviated with the Panther — only increased the hectic environment in planning and manufacturing. In addition, the Panzerkampfwagen E 50 from the "E-series" was eventually expected to replace the Panther II. The achievements of the industry, the Heereswaffenamt, the Reichsministerium für Rüstung und Kriegsproduktion and especially the troops were not impaired by such machinations. It must be stressed that despite nearly insurmountable technical problems, insufficient testing, rushed operations, and a shortage of raw and replacement materials an outstanding weapon was created in an inordinately short period of time, a weapon which continues to influence the course of international armor design in our day.

Appendixes

Appendices 1a and 1b

Biographical sketch of Ministerialrat Dipl.-Ing H.E. Kniepkamp

Ministerialrat a.D Dipl.-Ing Kniepkamp was born on January 5th, 1895 in Wuppertal. He participated in the First World War as a soldier with the Motorized Infantry and completed his degree at the Technishe Hochschule Karlsruhe. From March 1923 Kniepkamp worked as a gearbox design engineer at Maschinenfabrik Augsburg-Nürnberg, and following a short period with a Rostock engineering firm came to the Heereswaffenamt in Berlin in January of 1926.

There he primarily concerned himself with the design of wheeled/tracked, half-track and rapid-moving fully tracked chassis. As early as the end of 1926 contracts were awarded to the firm of Dürkopp for half-tracks, and to Krupp for wheeled/tracked vehicles. By the beginning of 1927 the first designs for 10 and 20 ton tanks had materialized, which were later tested in Russia. In spite of limited financial resources, in 1930 Kniepkamp created the first permanent testing facility for tracked vehicles.

In 1932 he drafted the first set of guidelines for design requirements pertaining to tanks and half-tracks, which led to the construction of the Panzerkampfwagen I and six different sizes of tractor design. In mid 1936 Kniepkamp was made Regierungsbaurat and was now responsible for the development of tracked vehicles.

The introduction of the torsion bar suspension, the interleaved running gear for tracked vehicles, the compact Maybach engines with their high power/weight ratio and finally the semi-automatic Maybach VARIOREX and OLVAR gearboxes were milestones of his involvement, which reached its zenith with the Tiger and Panther tanks. By the end of the war Kniepkamp had submitted approximately 50 patents in the area of tracked vehicles.

After the war's end Kniepkamp founded an engineering bureau, which quickly returned to the research and development of tracked vehicles. One of its better known designs was the running gear using Dubonet plate springs for the Swiss Pz 61 tank.

In 1957 two fully tracked armored personnel carrier prototypes were built based on his guidelines. Kniepkamp worked as an advisor from 1957 to 1960 on the development

Ministerialrat Dipl.-Ing. H.E. Kniepkamp.

of several prototypes for the German standard tank, which, like his personnel carriers were equipped with a hydro-pneumatic air suspension system.

Following 47 years (19 of them in the Heereswaffenamt) of devoted service in the area of tracked vehicles for military applications, on which he exerted a significant influence, Kniepkamp retired from development at the age of 78. He died on July 30th, 1977 in Heilbronn.

Biographical sketch of Professor Dr.-Ing. e.h. Karl Maybach

Professor Dr.-Ing e.h. Karl Maybach was born July 6th, 1879 in Cologne. After graduating from the Realschule he attended the Höhere Maschinenbauschule in Stuttgart and later the Polytechnikum in Lausanne. Upon completing his studies he went to work for the Daimler-Motoren-Gesellschaft in 1906. His outstanding abilities in the area of engine development for aircraft and airships, road and rail vehicles were recognized by an honorary doctorate from the Technische Hochschule Stuttgart.

As the German government resolved upon a comprehensive motorization of the army in the mid-1930s, it was the high running, high performance 6 and 12 cylinder engines of the firm of Maybach which were to be found in virtually all German tanks and prime movers. The pinnacle of this development was the 23 liter 700 hp HL 230 engine, which was installed in the Panther and Tiger tanks. By the war's end roughly 140,000 engines had been delivered — including 50,000 with disk-type crankshafts — for a total power output of approximately 40 million horsepower.

The Maybach company was also active in the area of transmissions. The semi-automatic pressure-driven VARIOREX gearbox was the first to appear for tractors and armored vehicles, later followed by the oil-pressure driven OLVAR transmission. With their potential eight gears using only four gear sets — due to the Maybach diverter dog clutch — and their preselector shifting, these transmissions were truly state of the art.

After the Second World War Karl Maybach created new types of high performance diesel engines based on the mechanical assembly technique. In the area of gearboxes, the continued development of the fully automatic hydromechanical "Mekydro" transmission was carried out for rail vehicles following tests for tank compatibility which had taken place near the end of the war. By the end of the 1950s the performance of diesel engines for naval application had increased to 3600 hp.

On 6 July 1959 Karl Maybach celebrated both his 80th birthday and the 50th anniversary of the firm of Maybach-

Professor Dr.-Ing e.h. Karl Maybach.

Motorenbau.

Karl Maybach closed his eyes for the last time on February 6th 1960. His life work continues today with the "Motoren- und Turbinen-Union", a company formed from the merger of MAN and Maybach-Mercedes-Benz.

Appendix 2

Panzerkampfwagen V Panther, (Sd Kfz 171) (VK 3002) and Panzerbefehlswagen Panther (Sd Kfz 267 and 268)
(compiled by Dokumentation Kraftfahrwesen e.V., Fachausschuss Wehrtechnik - Oberst a.D. Dipl.-Ing. Th. Icken

Developing firm: MAN, Nuremberg
Establishment of state of construction: 11 May 1942
Manufacturing firms: Daimler-Benz, Berlin-Marienfelde
Henschel, Kassel-Mittelfeld
MAN, Nuremberg
Maschinenfabrik Niedersachsen-Hannover (MNH), Hannover

Panzerkampfwagen Panther tanks completed to 31 July 1943

Daimler-Benz	202		1943	1783+50
Henschel	116 (200 by November)		1944	3768
			1945	441
MAN	209		total	6042
MNH	184			
Total	711			

note: during the Second World War tank production included the following:
Germany 1939-1945 total 25,000 - USA 49,000 Shermans; USSR 39,700 T 34s

Production: Pz Kpfw Panther, Pz Bef Wg Panther, Jagdpanther and Bergepanther

Production series	Ausf. D	chassis no. 210001-210254
		211001-216000
from September 1943	Ausf. A	chassis no. 210255-211000
		151000-160000
from March 1944	Ausf. G	chassis no. from 121301, from
		124301, and from 241001
	Ausf. F	beginning January 1945 at Daimler-Benz,
		Berlin-Marienfelde (not supplied to troops)

Panther losses (not counting Southern and Southeastern regions)

12/1/1943 to 6/30/1944(Eastern)	713
12/1/1943 to 9/30/1944(Eastern)	1303
12/1/1943 to 11/30/1944(Eastern)	2116
9/1/1944 to 11/30/1944(Western)	613

Batch costs from Heereswaffenamt WuG 6

Pz Kpfw Panther	117100 RM (minus gun)
Kanaone 7.5 cm KwK 42	12000 RM
HL 230 engine	11000 RM
AK 7-200 gearbox	3500 RM
MG 34	312 RM

Raw material requirements: iron, non-alloy 33409 kg
iron, alloy 44060 kg
iron, total 77469 kg

5 man crew

1. Weights and Dimensions

combat weight	kg	44800
combat weight with all-steel road wheels	kg	46500
drive		front
power to weight	kW/t(hp/t)	11.9(15.6)
specific ground pressure	bar	0.88
specific ground pressure at 200 mm soil sinkage	bar	0.7
turret weight with gun	kg	9500
gun with armor mantlet	kg	2650
MG 34	kg	11
engine(dry)	kg	1200
transmission(dry)	kg	1240
gearbox with clutch and bevel spur gear	kg	750
final drive	kg	395
cooling unit for engine(dry)	kg	200
drive sprocket	kg	275
idler wheel	kg	125
idler wheel arm	kg	92
road wheel (disk)	kg	75
all-rubber tires for road wheels	kg	18
swing arm	kg	59
1st and 7th swing arm	kg	72
track (86 links)	kg	2090
track link	kg	21
length(barrel at 12 o'clock)	m	8.66
length(barrel at 6 o'clock)	m	9.09
length of chassis	m	6.87
length of hull	m	6.60
barrel overhang at 12 o'clock	m	1.79
width(minus protective skirting)	m	3.27
width(with protective skirting)	m	3.42
track base	m	2.61
height with commander's cupola opened	m	3.10
height with commander's cupola closed	m	2.85
height with antennae	m	4.00
height of turret with commander's cupola	m	1.15
ground clearance	m	0.54
height of drive sprocket axle, front (measurement for vertical step)	m	0.826
hull length(exterior/interior)	m	6.56/6.40
hull width(exterior/interior)	m	1.84/1.76
hull height(exterior/interior)	m	1.32/1.29
volume	m³	7.26
width of track	m	0.66
ground contact length of track	m	3.92
ground contact length of track with 200 mm soil sinkage	m	4.92
steering ratio(track ground contact length: track base)	m	1.50

fuel supply	l(kg)	720(554)
fuel supply: combat weight	l/t	16
ratio of fighting compartment to total space	%	63.7
center of gravity, behind center of track ground contact	m	0.120
height of center of gravity above road wheel axles	m	0.650

2. Performance characteristics

maximum speed	kmh	55
speed with engine at 2500 r.p.m	kmh	46
maximum speed in 1st gear	kmh	4.1
cruising speed (paved surface)	kmh	33
cruising speed (terrain)	kmh	25
range on paved surface	km	200
range over terrain	km	100
fuel consumption (road)	l/100 km	350
fuel consumption (road)	l/t + 100 km	6.5
fuel consumption (medium rough terrain)	l/100 km	700
climb angle		30 degrees
climb angle in loose sand		26 degrees
vertical step	m	0.90
trench	m	1.90
fording	m	1.70
submersion (only Ausf. D and A)	m	4.00
maximum traction (on loose soil)	kN (t)	260(26)
kinetic energy at 45 kmh	mkN(mt)	3600 (36)

3. Main gun

type		7.5 cm KwK 42(L/70) 7.5 cm KwK 44/L (L/70) only Ausf. F
development and manufacturer, turret and gun		Rheimetall, Unterlüss
turret crew	men	3
muzzle height	m	2.26
caliber	mm	75
turret ring diameter	mm	1650
barrel length	mm	5250
barrel length with muzzle brake	mm	5535
load space length	mm	668
caliber length		70
weight of barrel, complete with breech and muzzle brake	kg	100
weight of gun, complete, with mantlet	kg	2650
smoke exhaust/muzzle brake		yes/yes
required gas pressure	bar	3200
rifling, number of grooves		32
rifling, depth of grooves	mm	0.9
width of grooves	mm	3.86+1/100
field width	mm	3.5+1/100
rifling, twist		6 degrees 30'
type of rifling	m	constant
recoil brake, liquid content	l	9
recoil length, normal	mm	400
recoil length, maximum (cessation of fire)	mm	430
barrel recuperator, liquid content	l	2.3
barrel recuperator, air pressure	bar	55+-5
compensator, liquid content	l	0.5
air pressure in compensator with fuse safety pin inserted	bar	75+-3
barrel blowout, air pressure	bar	10.5
maximum firing range	m	10000
turret bearing race, manufacturer		Kugelfischer und Vereinigte Kugellagerfabriken

3.1 Performance characteristics of ammunition

Ammunition type	Panzersprenggeschoss 39 PzGr 39/42	Panzergeschoss mit Stahlkern 40 PzGr 40/42	Sprenggranate 34
ammunition weight (cartridge) kg	14.30	11.55	11.14
shell weight kg	6.8	4.75	5.74
muzzle velocity V_o m/s	925	1120	700
muzzle kinetic energy mt	297	298	
cartridge length mm	893.2	875.2	929.2
no hollow charge ammunition			

3.1.1 Panzersprenggeschoss 39 Pz Gr 39/42

firing distance	shell velocity	penetrating force in mm (plate hardness 95-105 kg/mm²) at strike angle	
m	m/s	90 degrees	60 degrees
0	925	167	133
450	865	149	121
900	800	133	110
1350	730	118	99
1850	665	104	89
2300	605	91	79

3.1.2. Panzergeschoss mit Stahlkern 40 Pz/Gr 40/42

firing distance	shell velocity	penetrating force in mm (plate hardness 95-105 kg/mm²) at strike angle	
m	m/s	90 degrees	60 degrees
0	1120	230	197
450	1090	198	154
900	970	170	123
1350	865	145	99
1850	770	122	80
2300	675	103	65

Complete accounting of the gun's plotted hit firing tests against a vertical surface is as follows:

range	H	L
500	0.60	0.45
1000	0.80	0.50
1500	1.10	1.50
2000	1.50	1.75

After 2000-3000 rounds the gun barrel became worn and required changing. This meant:

Pz Gr 39	1 shot
Pz Gr 40	4 shots
SprGr 42	1/4 shots

After 500-700 rounds of Pz Gr 40 the gun barrel required changing.

The data in sections 3.1.1. and 3.1.2. were taken from the report "Le Panthe"" from the Ministère de la Guerre, Section Technique d l'Armeée, Groupement Auto-Chars, 1947.

firing range at 20 degrees elevation m	9850
rate of fire rounds/minute	6-8
ammunition supply piece	79 (Ausf. G 82, Pz Bef Wg Ausf. D 64), Pz Bef Wg Ausf. G 70)
total weight of ammunition (79 rounds)	1005 kg

Raw materials required for gun and accessories

Fe	3883 kg
Mo	1.4 kg
Cr	33.7 kg
W	0.12 kg
Sn	0.01 kg
Cu	0.6 kg
Al	0.06 kg
Pb	0.01 kg
Zn	1.65 kg
Ag	0.004 kg

4. Secondary weapon

Number of available machine guns	2 (1 for Pz Befehlswagen) only Ausf. F 1 (1 for Pz Befehlswagen)	
Type	MG 34 (Ausf. A 1)	MG 42
Caliber	7.9 mm	7.9 mm
Weight of gun	11 kg	10.6 kg radioman's MP 44 2 "P" attachments for compressed barrel
Length of gun	1225 m	1220 m
Length of barrel	0.60 m	0.53 m
Number of rifling	4	4
Muzzle velocity V_o	755 m/s	740 m/s
Rate of fire	900 rounds per min	1500 rounds per min
MG ammunition rounds carried		
Ausf. A and D	5100(34 satchels)	
Ausf. G	4800(32 satchels)	

5. Other weapons with data for available ammunition

26 mm Nahverteidigungswaffe in turret roof for firing smoke grenades (12 rounds), Sprenggranaten with time delay fuse (20 rounds), smoke markers and flares (24 rounds)

Machine pistol MP 40, caliber 9 mm, 1 gun

6. Gun control, targeting and vision systems

elevating system	mechanical
traverse system	hydraulic (15 bar pressure)
manufacturer of the gun control system	Böhringer
maximum traverse speed through 360 degrees(activated by foot pedal	60, Ausf. F 30
one turn of the hand crank alters the traverse by	0.365 degrees = 6.5-
one turn of the hand crank alters the elevation by	0.5 degrees = 8.9-(for Ausf. D)
range of elevation	Ausf. A and G -8 to +18 degrees Ausf. D and F -8 to +20 degrees
periscope type	TZF 12, binocular(Ausf. D) TZF 12a, monocular(Ausf. A and G) TZF 13, binocular(Ausf. F) stabilized periscope (supplementary)
manufacturer	Leitz, Wetzlar
magnification	x 2.5 and 5
field-of-view	at 2.5x magnification 28 degrees = 498- at 5x magnification 14 degrees = 249-
exit pupils diameter mm	6 and 3
optical length mm	1140
movement range of the articulated periscope TZF 12	+30 to -20 degrees
articulated periscopeTZF 12a	+40 to -30 degrees

Panther Ausf.	D	A	G	F
glass blocks (6 piece) manufactured by Sigla commander's cupola fixed mirror periscopes (7 pieces)	yes			
commander's cupola		yes	yes	yes
periscope	driver 1 R/O 1	driver 1 R/O 1 loader 1	R/O 1 loader 1	R/O 1
driver's periscope, rotatable Manufacturer: Meyer, Görlitz			driver 1	driver 1

gunner's ball sight	KZF 2 with 1.8x magnification	
	horizontal range	+_5 degrees
	vertical range	-10 degrees to +15 degrees

Horizontal stereoscopic rangefinder with 1.32 m base and 15x magnification was used in the Befehlspanzer and tested in the Ausf. F (developed by Zeiss, Jena)

7. Communications equipment

	equipment type	frequency range in MHz	average maximum range in km	voice signal
PzKpfw Panther	Fu 5 10 W transmitter "c" with UHF receiver "e"	27-33	6.4	9.5
	Fu 2	27-33		
Pz Befehls Wg	Fu 8			
Sd Kfz 267 (Pz Brigade/Regiment to Pz Battalion)	30 W transmitter	0.83-3.024		80
	HF receiver "c"	0.83-3.0		
	Fu 5 10 W transmitter	27-33	6.4	9.6
	VHF receiver	27-33		
Pz Befehls Wg Sd Kfz 268 (Pz Battalion to Pz Company or individual tanks)	Fu 7 20 W transmitter "c" with UHF receiver	27-33	12.8	16.0
		27-33		
	Fu 5 10 W transmitter	27-33	6.4	9.6
	UHF receiver "e"	27-33		

8. HL 230 engine

Technical data for HL 210 in parentheses

Development	Maybach Motorenbau, Friedrichshafen
production	Maybach Motorenbau, Friedrichshafen Nordbau, Norddeutsche Motorenbau GmbH Berlin-Niederschöneweide (assembly plant) Auto-Union Werk Wanderer, Chemnitz Daimler-Benz, Stuttgert-Untertükheim
Type	HL 230 P30
Number of HL 230 engines completed for	Panther 7000 Tiger 2000 (Type HL 230 P45) total 9000

Engine housing was cast grey steel with lubricated cylinder sleeves (for series production)

Pistons	Mahle, Stuttgart-Bad Cannstatt with lead bronze coated interchangeable connecting rod steel bearing bushings		
Operating method	4-stroke gasoline-carburetor		
Number of cylinders/ arrangement	12/60 degrees V		
Bore/stroke	mm	130/145 (125/145)	
swept volume	l	23.0 (21.35)	
Compression ratio		6.8:1	
Octane rating		74	
Maximum output at rpm	kW(hp) 1/min	515(700)/3000	
Maximum torque at rpm	mN (mkg) 1/min	1850(185)/ 2100	
at idle rpm	1/min	800	
Combustion air filter	combination filter (cyclone dry filter and oil bath filter) Mann u. Hummel, Ludwigsburg		
Cooling	liquid		
Quantity (entire cooling system)	l	170	
Max. fan rpm (with two-speed motor)	1/min	3000/4500	
Max. coolant temperature	degrees Centigrade	90	
Volume of air drawn in	kg/hp/h	3.4	
Quantity of engine lubrication			
initial fillup	l	42	
oil change	l	32	
Specific fuel consumption	g/hp/h	260	
Intake valve opens	10 degrees before O.T.		
Intake valve closes	50 degrees after U.T.		
Exhaust valve opens	45 degrees before U.T.		
Exhaust valve closes	15 degrees after O.T.		
Piston play	mm	0.14-0.16	
Valve play (intake and exhaust) with engine cold	mm	0.35	
Carburetors	4 Solex double downdraft cross-country carburetors Type 52 FF J II D (2-stage carburetor)		
Weight of engine in cast grey steel form (dry)	kg	1200	
Power/weight ratio	kg/hp	1.72	
Output per unit of swept volume	hp/l	30.5	
Width of engine	mm	1000	
Height of engine	mm	1190	
Length of engine	mm	1310	
Installed engine volume	m^3	1.56	
power-to-space ratio	hp/m^3	4.48	
Volume of engine compartment	m^3	4.2	
Engine compartment space utilization	hp/m^3	166	
Piston surface performance N/F	hp/cm^2	14.5	
Average piston pressure	bar	9.1	
Average piston speed	m/s	14.5	
Cooling system, max. permissible external temperature degrees	Centigrade	+42	
Production man hours for completing one engine		1260	
Iron and steel composition, non-alloy, per engine	kg	2100	
alloy	kg	560	

make	model	gasoline/diesel	stroke	liquid/air	type	bore(mm)	HUB(mm)	volume	rpm	performance (hp)
BMW	132 Dc	gasoline	4	air	9*	155.5	162	27700	2000	650
Maybach	HL 234	gasoline	4	liquid	12 V	130	145	23000	3000	850
F.K.F.S.	(Kamm)	gasoline	4	air	2x12 v	135	140	48000	2050	1000
Daimler-Benz	MB 507	diesel	4	liquid	12 V	158	180	42300	2300	850
Deutz	T8 M118	diesel	2	liquid	8 V	170	180	32300	2000	700
Deutz	Dz 710	diesel	2	liquid	16-.-	160	160	51500	2500	1500
MAN/Argus	LD 220	diesel	4	air	16 H	135	165	37800	2200	700
Simmering	Sla. 16	diesel	4	air	16 X	135	160	36500	2000	720

V=V engine *=radial engine H=H engine X=X engine -.-=Boxer engine

9. Transmission, clutch

Drive front

9.1 Gearbox

development	Zahnradfabrik Friedrichshafen
	ZF Friedrichshafen
	production ZF Friedrichshafen
	Waldwerke Passau
	(assembly plant)
	Steyr-Daimler-Puch,
	Graz-Thorndorf
	(to Dec. 1942 500 units produced)
	Lanz, Mannheim
model	AK 7-200
type of gearbox	all-synchromesh, 2nd-7th gear
synchronized	synchronized
number of gears(forward/reverse)	7/1
entire reduction ratio	1:13.4
weight including main clutch and bevel gear kg	750

Speed, turning radius and tractive power

	reduction ratio	jump	speed in km/h at rpm 2500 1/min	3000 1/min	turning radius m	tractive power at 2500 rpm t
1st gear	9.21	2.02	3.5	4.1	5	50
2nd gear	4.56	1.59	7	8.4	11	25
3rd gear	2.87	1.56	11	13.3	18	15
4th gear	1.83	1.45	18	20.8	30	10
5th gear	1.27	1.41	25	30.8	43	7
6th gear	0.90	1.31	35	42.5	61	5
7th gear	0.69		46	55.0	80	4
reverse	9.46		3.4	4.0	5	50

9.2 Main clutch

type	triple plate dry clutch
model	LAG 3/70 H
max. rpm mkg	200
manufacturer	Fichtel & Sachs, Schweinfurt

9.3 Steering unit

developer	MAN, Nuremberg
type	single-radius controlled differential discontinuous
regenerative	
main bevel drive, reduction ratio and number of teeth	1:1.05(21:20)
planetary gearing(compound drive), reduction ratio and number of teeth	1:1.4 60+24/60*
steering drive, reduction ratio and number of teeth	1:1.95 (35:18)
spur gear drive, reduction ratio and number of teeth	1:4.65 (79:17)
steering clutch	single-radius dry clutch, hydraulically operated
support brakes	external contracting brake with cast iron coating, mechanically operated with aid of hydraulic pressure
steering brakes	solid disk brakes, type LG 90, mechanically operated with aid of hydraulic pressure 60 bar, manufacturer Argus, developer Dr.-Ing Herman Klaue
foot and hand brake	operate both steering brakes

9.4 Final drive

reduction ratio and number of teeth of spur gear	1:8.4 (spur gears 24:11 and 38:10)

10. Running gear

type of running gear	interleaved, Ausf. F staggered
type of suspension	double torsion bar, hairpin design (two torsion bars per road wheel)
developer	MAN (Dr.-Ing habil. Ernst Lehr)
manufacturer of the torsion bars	Dittmann-Neuhaus,
Herbede	(Ruhr); Hoesch, Hoehenlimburg(Westphalia); Röchling, Völkingen(Saar)
travel stroke (road wheel vertical movement)	mm 510 (compare to Tiger 220, T 34 240, Sherman 111, Leopard 1 373, Leopard 2 530)
shock absorbers per side	2 Hemscheidt hydraulic shock absorbers, type HT 90, operating independently, on 2nd and 7th swing arm
length of swing arm, wrought iron, manufactured by Siepmann, Delecke (Möhnesee)	mm 420
angle of swing arm(static) offset from horizontal	25 degrees
angle of swing arm movement from its lowest position to horizontal	61 degrees
road wheels per side	8 (4 inner and 4 outer)
static road wheel pressure	kg 2500
average load per road wheel width N/cm (kg/cm)	1410 (141)
type of road wheels	disked, all-rubber tires

	(removeable) during Ausf. G production run, all-steel resilient wheels introduced (manufactured by Deutsche Eisenwerke, Mühlheim-Ruhr)
interval of road wheel axles	mm 560
road wheel diameter	mm 860
road wheel width	mm 100
return rollers per side	1
protective skirting, plate thickness	mm 5
drive sprocket, number of teeth	17
drive sprocket pitch diameter	mm 821.77
track type	hinged, dry pin, with mounts for snow grousers on every 5th to 7th track link
track model	gs 64/660/150
track developer	Moorburger Treckerwerke Hamburg-Moorburg
number of links	86
non-skid pads for every 5th and 7th track link(max. speed 15 km/h)	
pitch	mm 150
width	mm 660
sprung weight	t 39.9
sprung mass	kg s²/cm 40.7
mass moment of inertia around the center of gravity	mkg s² 14.4×10^3
height of center of gravity above road wheel axles	m 0.65
spring constant on road wheel (with static load)	kg/mm 8 (compare Tiger 1st and 9th road wheel 31.3, 2nd-8th road wheel 26; T 34 20, Sherman 16, Leopard 1 24, Leopard 2 21)
natural frequency of bounce	1/min 51.5
natural frequency of pitch	1/min 31 (30-50 targeted)
damping of bounce	25%
damping of pitch %	40 (30-40 targeted)

11. Electrical systems

generator	Bosch 0.7 kW 12 Volt DC, Model GTLN 700/12-1500 Bl 1 (HL 210, 1.0 kW)
voltage regulator	Bosch Model RS/KN 600/12/1 or SSM 41 L 21 Z
electric starter	Bosch 6 hp, 24 Volt, Model BPD 6/24 ARS 15
inertia starter	Bosch Model AL/ZM 1
batteries	2x12 V 150 Ah Model 12 B 150 or 2x12 V 120 Ah Model 12 B 120 Pz (heatable using heating plates with two settings of 100 and 300 Watt)
ignition	2 magnetos with impulse starter installed, Model JGN 6/R 18
contact breaker interval	mm 0.35
spark timing	5 degrees after O.T.
spark adjustment	automatic
firing order	12-1-8-5-10-3-7-6-11-2-9-4
spark plugs	Bosch W 225 T 1
interference suppression	Bosch 2xEM/S 75/1 or 1xEM/S 100/1 or 1xEM/S 5/1

12. Armor protection

manufacturer of hull and turret housing	Dortmunder-Hoerder-Hüttenverein, Dortmund; Ruhrstahl, Hattingen; Bismarckhütte, Upper Silesia; Harkort-Eicken, Hagen (Westphalia) Eisenwerke Oberdona, Linz (Upper Austria); Böhler, Kapfenberg und Deuchendorf (Steiermark); Schoeller-Bleckman, Mürzzuschlag (Steiermark)
thickness, slope and equivalent thickness	
nose, lower mm/degrees/mm	60/35/105
nose, upper	80/35/140
side, upper	40/50/52 (Ausf. G 50/60/58)
side, lower	40/90/40
rear	40/60/46
bottom	front 26/0, rear 17/0
engine compartment cover	15/0
turret mantlet	120
turret front	110/78/112 (D 80/78/82, F 120/80/122, G 100/80/102)
turret rear	45/65/50 (F 60/65/66)
turret side	45/65/50 (F 60/65/66)
turret roof	17/5 (F and G 30/5)
track shields(skirting)	5/90/5
floor, fighting compartment/ engine compartment	25/16

driver and radio operator entry hatches swivel to the side, balanced on springs (Ausf. A and D)
driver and radio operator entry hatches opening upward (Ausf. G)

13. Fluid fill quantities

total for cooling system	l	120
type of engine oil:		Motorenöl der Wehrmacht
engine at initial filling	l	42
engine at oil change	l	32
2 x fan drives	l	1.5
combination air filter, each	l	0.75
type of transmission fluid:		Getriebeöl der Wehrmacht
gearbox	l	33
gearbox, fluid reservoir	l	12
gearbox, housing	l	21
oil cooler with lines	l	15
clutch bearings	l	0.3
final drive, each	l	5.5
turret drive	l	2.75
type of fluid:		Stossdämperöl der Wehrmacht nach TL 6027
hydraulic pressure unit(brakes)	l	7
shock absorbers, each	l	1.75
lubricating oil		Einheitsabschmierfett der Wehrmacht
running gear (at initial filling)	kg	38
fuel		74 octane (OZ 74)
5 fuel tanks, totalling	l	720
fill tank	l	65
tank, upper left	l	210

tank, lower left	l	110	
tank, upper right	l	210	
tank, lower right	l	125	

Roughly beginning with the Ausf. G the left tank group was separated from the right; each held 360 and 370 liters, respectively.

14. Miscellaneous

Coolant pre-heating	utilized (Fuchs-Gerät, heating by means of a heater lamp)
Battery pre-heating (Bosch)	100 and 300 Watt (heating plates)
Biological/chemical protection (Dräger)	under testing
Fighting compartment heating	from fans of the cooling system
Floor escape hatches	not available

Automatic fire extinguishing system with 3 heat sensors, 3 extinguisher nozzles and a 3 liter bottle with 2 liters of CB extinguisher (Minimax)
Gyrocompass Modell B *Anschütz used
Submersion equipment only for specialized units

Appendix 3

Jagdpanther (Sd Kfz 173)
"8.8 cm PaK 43/3 auf Panzerjäger Panther"

(data only provided when different than that of Panzerkampfwagen Panther)
(compiled by Dokumentation Kraftfahrwesen e.V., Fachausschuss Wehrtechnik - Oberst a.D. Dipl.-Ing Th. Icken)

developing firm	Krupp, Essen		
	Daimler-Benz, Berlin-Marienfelde		
	Mühlenbau und Industrie AG (MIAG), Braunschweig		
initial demonstration	20 October 1943 at Arys troop training grounds (East Prussia)		
manufacturer	Mühlenbau und Industrie AG (MIAG), Braunschweig from February 1944		
	Maschinenfabrik-Niedersachsen-Hannover MNH), Hannover, from December 1944		
number completed	384	1944	212
		1945	172
chassis numbers	300001-300392		
raw material requirements	iron, non-alloy	34425 kg	
	iron, alloy	41327	
	iron, total	75752	
cost of gun	21000 RM		
number of crew	5		

1. Weights and dimensions

combat weight	t	45.5
power-to-weight ratio	kW/t(hp/t)	11.6(15.4)
specific ground pressure	bar	0.87
length	m	9.87
length of chassis	m	6.87
barrel overhang	m	3.00
overall height	m	2.72
muzzle height	m	1.96

2. Main gun

development		Krupp, Essen
manufacturer		Dortmunder Hoerder Hüttenverein, Werk Lippstadt
type		with welded collar mantlet 8.8 cm Pak 43/3 (L/71) with bolted collar mantlet 8.8 cm Pak 43/4 (L/71)
muzzle height	m	1.96
caliber	mm	88
barrel length with muzzle brake	mm	6686
caliber length		71
distance from rear baseplate surface to throat at forward breech recess surface	mm	290
distance of bore from throat at forward breech recess to muzzle	mm	6010
length of rifled part	mm	5150.5
length of rifled part in calibers		58.5

Rifling:

number		32
depth	mm	1.2

width	mm	5.04+0.6	
field width	mm	3.6-0.6	

Chamber:
diameter of rear tapered part			
rear	mm	132.4	
front	mm	123.9	
diameter of forward tapered part(forcing cone)			
rear	mm	92.5	
front	mm	88.0	
length of chamber	mm	859.5	
rifling, twist		6 degrees 30' (27.57 caliber)	

Weights:
barrel, complete with breechblock and muzzle brake	kg	1690
barrel, complete with breechblock	kg	1628
barrel, complete	kg	1225
baseplate minus breechblock	kg	275
chuck	kg	26
wedge breechblock with inner parts	kg	50
wedge breechblock with operating assembly	kg	70
muzzle brake	kg	62
recoil brake	kg	65
barrel recuperator	kg	50
total weight of gun	kg	2200

Performance characteristics:
required gas pressure	bar	3000
design gas pressure	bar	3700

Recoil brake:
average braking force	kg	6300
fluid content	l	5.4
recoil length, normal	mm	550
recoil length, (max) "cease fire" mm		580

Barrel recuperator:
initial compression of air	bar	50
fluid content	l	5.3
firing		electric

Raw material requirements for one gun in kg
Fe	4421
Mo	8.1
Cr	48.4
V	4.85
Sn	0.02
Cu	0.126
Ni	29.5
rubber	1.1

Performance characteristics of ammunition

ammunition type	Panzer-spreng-geschoss 39 PzGr 39/43	Panzer-geschoss mit Stahl-kern 40 PzGr 40/43 W	Spreng-granate 43	Hohlla-dungs granate HL GR 39
ammunition weight (cartridge) kg	23.4	19.9	18.7	15.35
shell weight kg	10.2	7.3	9.4	7.65
weight of propellant charge kg	6.8	6.8	3.8	2.0
explosive charge kg	0.05	–	1.0	0.77
muzzle velocity m/s	1000	1130	750	600
muzzle kinetic energy mt	516	480	269	140
cartridge length mm	1124	1124	1170	

penetrating force at 90-degree entry angle in mm at distance in meters(plate hardness 95-105 kg/mm2) The data in parentheses refer to a 60-degree entry angle

100 m	220(203)	300(250)		90
500 m	205(182)	270(225)		90
1000 m	186(167)	237(190)		90
1500 m	170(150)	205(160)		90
2000 m	154(135)	175(135)		90

A significant improvement in the penetrating force of current tank guns is, as of this time, only shown in the area of hollow charge(Hohlladung) rounds

range at 15 degrees elevation m	9350
rate of fire rounds/min	6-8
ammunition carried each	57
ammunition weight kg	1226

3. Secondary weapon
number of available machine guns	1(nose)
type	MG 34
machine gun ammunition carried rounds	600

4. Other weaponry

1 close-combat weapon(roof plate) 26 mm
 2 machine pistols, caliber MP 40 9 mm (384 rounds)

5. Gun control, targeting and vision systems
Sighting equipment - telescopic periscope 1/4
2 periscopes each for commander, loader and driver
elevation range	-8 to +14 degrees
traverse range	+-11 degrees

Targeting system
marking of angle-of-elevation dial
for Pz Gr 39/43	m	0-4000
for Spgr 43	m	0-5400
for Gr 39 HL	m	0-3000
for Pz Gr 40/3	m	0-4000

graduation, fine	0-100
graduation, coarse	0-300
angle of site graduation, fine	0-100
angle of site graduation, coarse	100-500

Appendix 4

Designations for tank tracks

From the Heereswaffenamt/WaPrüf 6 the following designations were officially introduced for tracks:

I. Construction type:
1. rapid tracks for vehicles (as opposed to agricultural tractors) = K
2. six wheel tracks for multiple axle driven wheeled vehicles = S
3. tracks for half-tracked vehicles = Z
4. tracks for agricultural tractors = L
5. tracks for test purposes (test track) = P

II. Material used in track links
1. cast steel all alloys = g
2. steel, drop forged (pressed steel) = p
3. steel plate = b
4. tempered cast = t
5. Duraluminium = d
6. weaved = ge
7. silumin = s

III. Track link connections
1. normal fastening (bolts and sleeves no lubrication) = no symbol
2. floating pins = s
3. roller bearing (lubricated and sealed) = w 4. rubber hinged = gu
5. floating sleeves = b

IV. Design
V. Width in Arabic numerals
VI. Pitch
For example, the Panther track "Kgs 64/660/150" =
K = rapid tracks for vehicles
g = cast steel all alloys
s = floating pins
64 = design of track
660 = width of track in mm
150 = pitch of track in mm

Note: for registration in the Heereswaffenamt, behind the construction type there might have also been the following designations:
e = German-made track
f = foreign-made track

Appendix 5

Surface treatment of torsion bars*

The effect of increasing the life of springs in general applies to a strengthening of the surface which is only effective to a depth of just a few tenths of a millimeter. During the Second World War two different methods were employed, namely:
— the shot peening, or cloudburst hardening method based on the Röchling patet DRP 573 630 KL 18 C from 22 May 1929 and
— the "Rollverdichten" (roller compression) process according to a patent by Professor Otto Föppl (Technische Hochschule Braunschweig), DRP 521 405 from 10 July 1928.

Shot peening was used by the firms of Röchling and Hoesch. Initially steel grains comprised of miniature cast steel balls (diameter ca. 1 mm) were used as the shot material; later this was made up of particles of piano wire, cut into the same length as its diameter (so-called cutwire). The particles were "blown" at a high velocity onto the surface of the material being treated using fan blowers, which resulted in the previously mentioned effect of cloudburst hardening.

Dittmann & Neuhaus achieved the same effect using roller hardening. This process involved using a discus shaped roller having a small canted edge being placed under high pressure contact against the shaft of the torsion bar. Both the axis of the roller and that of the torsion bar were nearly parallel, and the roller was slowly drawn lengthwise along the rapidly spinning torsion bar so that the entire surface of the bar was subject to the hertzian face pressure, although only for a short time on each point of the surface.

* Provided by Dir. Dipl.-Ing Gerhard Geissler, Hoesch Werke on 14 December 1976.

Appendix 6: Technical data

Vehicle designator	Panzerkampfwagen V	Panzerkampfwagen V	Panzerkampfwagen Panther (Sd.Kfz.171)
Model			D
Type	VK 3002(DB)	VK 3002(MAN)	VK 3002(MAN)
Manufacturer	Daimler-Benz	MAN	MAN/HS/DB/MNH
Year	1941-1942	1941-1942	1942-1943
Information source	DB, archives	MANZeichnung Tu 16901	Handbuch WaA, Blatt K 49
Engine			
Manufacturer, model	Daimler-Benz "MB 507"	Maybach "HL 210 P 45"	Maybach "HL 230 P 30"
Number of cylinders, arrangement	12, V-form 60 deg.	12, V-form 60 deg.	12, V-form 60 deg.
Bore/stroke(mm)	162/180	125/145	130/145
Swept volume(cm^3)	44500	21353	23095
Compression ratio	14.8	7	6.8
Rpm: (normal/maximum)	2000	3000	2500/3000
Maximum output(hp)	650	650	600/700
Power-to-weight(hp/t)	19	20	15.5
Valve arrangement	overhead	overhead	overhead
Crankshaft bearings	7 sleeve	7 roller	7 roller
Carburetor/fuel injection	diesel-2 Bosch PE 6	4 Solex 52 JFF II D	4 Solex 52 JFF II D
Firing sequence	1-11-2-9-4-7-6-8-5-10-3-12	12-1-8-5-10-3-7-6-11-2-9-4	12-1-8-5-10-3-7-6-11-2-9-4
Starter	Bosch	Bosch BPD 6/24 ARS 150	Bosch BPD 6/24 ARS 150 + Bosch AL/RBI/R 1 or AL/ZMI
Generator	Bosch	Bosch GTLN 700/12-1500	Bosch GTLN 700/12-1500 BL 1
Battery: number/Volt/Ah	6/12/120	2/12/150	2/12/150 and 120
Fuel delivery	pumps	2 mechanical pumps	2 mechanical pumps
Cooling	liquid	liquid	liquid
Clutch	Ortlinghaus hydr. multi-layered	triple plate, dry	triple plate, dry F&S LAG 3/70 H
Gearbox	DB "KSG 8-200"	ZF "AK 7-200"	ZF "AK 7-200"
Number of gears:forward/reverse	8/1	7/1	7/1
Track drive sprockets	rear	front	front
Final drive reduction ratio	4.8:1	8.4:1	8.4:1
Maximum speed (km/h)	56	55	55 road, 30 cross-country
Range: road/cross-country (km)	195/140	270/195	250/100
Steering type		MAN clutch and brake	MAN single-radius controlled differential
Turning radius ø (m)	in place	10.0	10.0
Suspension	leaf springs, lengthwise	torsion bars, crosswise	torsion bars, crosswise in double arrangement
Chassis lubrication system	high pressure and central	high pressure and lube batteries	high pressure and 4 cent. batteries
Braking system: manufacturer	Daimler-Benz	Süddeutsche Arguswerke	Süddeutsche Arguswerke
operating method	mechanical, hydraulic backup	hydraulic	mechanical, hydraulic backup
brake type	internal expanding	solid disk	solid disk, Model LB 900.2
Braking system operates on	drive unit	steering brake	steering brake
Type of running gear	interleaved	interleaved	interleaved
Type and size of road wheels	900x100, ø 900 mm	860/100, ø 860 mm	860/100 D, ø 860 mm
Track base (mm)	2740	2610	2610
Track ground contact length (mm)	3838	3920	3920/4920 with 20 cm soil sinkage
Track width (mm)	530	660	660
Track type		Kgs 64/660/150	Kgs 64/660/150
Number of links per track			86
Ground clearance (mm)	500	500	560
Overall length (mm)	6000	8625	-8660/9090-6870
Overall width (mm)	3200	3270	3270, with skirts 3420
Overall height (mm)	2690	2867	2995
Ground pressure (bar)		0.73	0.88
Combat weight (kg)	34000	35000	44800
Crew	5	5	5
Fuel consumption (l/100 km)		road 350, cross-country 700	road 280/cross-country 700
Fuel supply (l)	550	750	720
Armor of hull front (mm)	60	60	80
side (mm)	40	40	40
rear (mm)	50	40	40
turret front (mm)	100	80	110
side (mm)	40	45	45
rear (mm)	40	45	45
Performance climb (degrees)	30	30	30
vertical step (mm)	900	900	900
ford (mm)		1900	1900
trench (mm)		2450	
Armament, main*)	1 7.5 cm KwK 42 L/70( )	1 7.5 cm KwK 42 L/70( )	1 7.5 cm KwK 42 L/70(79)
auxiliary	2 MG 34	2 MG 34	2 MG 34(4200)

Comments *) data in ( ) shows ammunition carried

*)prototypes with "HL 210 P30" engine and clutch-and-brake steering

Panzerkampfwagen Panther (Sd.Kfz.171) A	Panzerkampfwagen Panther (Sd.Kfz.171) G	Panzerbergewagen Bergepanther (Sd.Kfz.179) A and G	Panzerjäger Panther (Jagdpanther) (Sd.Kfz.173)
VK 3002(MAN)	VK 3002(MAN)	VK 3002(MAN)	VK 3002(MAN)
MAN/DB/MNH	MAN/DB/MNH	MAN, HS, Demag	MIAG, MNH
1943-1944	1944-1945	1943-1945	1944-1945
D 655/1a from 21 Sep 1944	D 655/60 from 1 Nov 1944	D 655/1a	from 21 Sep 1944 D 655/60 from 1 Nov 1944
Maybach "HL 230 P 30"			
12, V-form 60 deg.			
130/145			
23095			
6.8			
2500/3000			
600/700			
15.5			
overhead			
7 roller		7+1 roller	7+1 roller
4 Solex 52 JFF II D			
12-1-8-5-10-3-7-6-11-2-9-4			
Bosch BPD 6/24 ARS 150 +			
Bosch AL/RBI/R 1 or AL/ZMI			
Bosch GTLN 700/12-1500 BL 1			
2/12/150 and 120			
2 mechanical pumps			
liquid			
triple plate, dry F&S LAG 3/70 H			
ZF "AK 7-200"			
7/1			
front			
8.4:1			
55 road, 30 cross-country		55/45.7	
250/100			250/100
320/160			
MAN single-radius controlled differential			
10.0			
torsion bars, crosswise in double arrangement			
high pressure and 4 cent. batteries			
Süddeutsche Arguswerke			
mechanical, hydraulic backup			
solid disk, Model LB 900.2			
steering brake			
interleaved			
860/100 D, ø 860 mm			
2610			
3920/4920 with 20 cm soil sinkage			
660			
Kgs 64/660/150			
86			
560			
-866-/9090-6870		8860	9870
3270, with skirts 3420		3270, with skirts 3420	3270, with skirts 3420
2995		2700	2715
0.88		0.83	0.89
44800		4300	45500
5		3	5
road 280/cross-country 700		road 280/cross country 700	road 280/cross-country 700
720		1075	720
80		80	80
40		40	45
40		40	40
110		–	–
45		–	–
45		–	–
30		–	–
900		900	900
1900		1900	1550
2450		2450	–
1 7.5 cm KwK 42 L/70(79), Ausf. G(82)		1 2 cm KwK 38	1 8.8 cm Pak 43/3 L71(57)
2 MG 34(4200)		–	1 MG 34(600)

Appendix 7

Data for Panther production (MAN)

It took a total of 2000 working hours to produce a single complete Panther. This included the following:
— Hull production: 55 hours
— Turret production: 38 hours
— Chassis assembly: 485 hours
— Turret assembly: 150 hours
— Final assembly: 85 hours

The following specialized machinery was used in production:
— three 8-spindle horizontal borers
— one vertical two-tool lathe
— eight Wesselmann drilling machines
— one radial boring machine
— one Heller double horizontal end milling machine,
— one turret lathe

No type of specialized machinery was necessary for assembly.

The 8-spindle horizontal borer was used for making the holes in the tank hull sides for the swing arms and torsion bars.

The vertical lathe formed the borings and seat for the turret gear ring and bearing race.

The Wesselmann drilling machines bored the openings for the shock absorber bolts.

The radial boring machine created the borings for anchoring the turret race ring.

The Heller milling machine was used for milling the hull side plates to receive the final drive.

The vertical turret lathe was utilized in finishing the openings for the turret ring and commander's cupola.

MAN, Nuremberg, manufactured the 8-spindle horizontal borer, the vertical lathe and the radial borer. The drilling machines were supplied by Wesselmann in Gera, the milling machine by Gebrüder Heller, Nürtingen, and the turret lathe by Droop & Rein, Bielefeld.

A total of 9968 workers were employed in the Maschinenfabrik Augsburg-Nürnberg AG works on 1 March 1945, of which 5448 were involved in tank construction. These were broken down as follows:
— in administration: 124
— in tank machining dept.: 841
— in tank manufacturing: 3983
— in tank assembly: 500

5023 of these were men, 425 were women. 2719 of the men were foreigners; among the women 230 were non-German.

Two shifts were run within a 24-hour period, each shift comprising 12 hours.

Appendix 8

Combat report of a Panther battalion

The latest operations of the unit, during which approximately 30 Panthers were in constant combat over a period of six consecutive days, have solidified the outstanding capabilities of the Panther tank. With a combination of well-trained personnel, careful handling and tactically sound operations great victories are possible. During these six days the battalion was able to destroy 89 tanks and assault guns, 150 guns, anti-tank guns, anti-aircraft guns, etc.

Despite the massed enemy defenses only 6 vehicles were put completely out of commission by enemy fire. The following lessons were learned during the operation: the high firing performance of the gun is to be exploited in all circumstances. Firing can commence at a range of 2000 meters with very favorable results. The bulk of all heavy weapons and tanks which were destroyed were done so at a combat range of between 1400-2000 meters. The ammunition required was relatively low; every fourth to fifth shot was a direct hit, even when using HE shells.

The attack profiles which had hitherto been planned, such as the wedge and inverted wedge — with their corresponding distances and intervals — should not be employed with the Panther. Distance and interval should be doubled for the wedge and inverted wedge profiles. Teamwork in pairs should be given greater emphasis within the group.

In all instances, the enemy should be denied the opportunity of engaging the Panther at closer ranges. Wide-ranging battlefield reconnaissance is therefore indispensable. It has proven necessary to send a platoon of Panther tanks 1000-1200 meters ahead, drawing the enemy fire out early so that the remaining tanks can open fire at a safe distance.

Flank protection is to be given to the Panther tank due to its sensitive side armor. In all cases, the unit leader should hold a reserve of tanks back with which he can snuff out any threats to his flank at a moment's notice. Pulling back forward elements is fundamentally a belated tactic. The fighting reserve should generally be held back at a distance of 1000 meters. It has proven necessary at the regimental level that the available Panzerkampfwagen IV tanks assume the role of flank protection, while the Panther tanks rush forward as a spearhead and threaten the enemy. Concentrated operations, which was generally possible within the battalion, led to major shocks to morale among the enemy during the fighting east of A. Multiple fire concentrations were conducted against certain locations, leading to the Russians fleeing their heavy weapons, and it was also noted that crews abandoned their T-34 tanks without their vehicles even being hit. During these types of fire concentrations a mixture of HE and armor piercing shells should be employed. The tactical commander should direct the number of ammunition rounds to be fired per vehicle by radio. Based on prisoner interrogation it was found that the Russian is extremely impressed with the flat trajectory of the 75mm KwK L/70 and will avoid engaging the Panther tank in open combat when he does not have a superior number of tanks.

In spite of the improved engine performance (the battalion had reached an average of 700 km per tank, with only 11 engine changes), it is basically recommended that the tanks be loaded when travelling distances of over 100 km — since the running gear suffers tremendously, especially in winter.

Conclusion:

The Generalinspekteur der Panzertruppen agrees entirely with every point in the battalion report.

A number of guidelines were contained in the report which are absolutely necessary for the success of Panther unit operations. In particular, the following points are critical:

1. Extended attack profiles. The report mentions intervals, which should be doubled.
2. Panthers keep the enemy from getting in close, since their strength is in long-range firing (2500 m and greater). For this a timely deployment of reconnaissance is needed. Combat reconnaissance in tank units often does not extend out far enough. Nevertheless, it can play a decisive role. This applies to all types of tanks.
3. Securing the flanks of the Panther through Panzer IVs, Panzergrenadiere and self-propelled guns.
4. At long last the battalion was employed as a single entity, contrary to the often-used traditional methods. Success lies at the heart of these types of operation. In and of itself it offers rapid and decisive results.
5. Concentrated fire against important targets. This guideline has also been incorporated into the new shooting directive, paragraph 189/192; up to this point it has seldom been used, yet offers success of great proportions.
6. Given the currently limited fatigue life of the tank engine, every opportunity should be taken to make use of rail transportation.

Appendix 9

Kraftfahrversuchsstelle des Heereswaffenamtes Prüfwesen
Aktz.: 76a/m 14/43/L Kummersdorf-Schiessplatz, 13 July
1943 J 77-127

re: Comparison of single-radius and dual-radius steering

a) Compared to the dual-radius steering of the Tiger, the single-radius steering as used in the Panther is a compromise solution fraught with serious drawbacks. The fixed radius in the production version is too small in quite a few cases, and in some too large.

Since the turning radius effected by decoupling (activating the support brakes) is often quite inadequate for achieving the necessary steering movement, driving must be done by either slip-clutching the fixed radius, or by skipping this stage using track braking. The clutch, however, isn't designed for slip-clutch loading, causing the wear to become intolerably great. If it is intended to bypass this step in order to use the track brakes, a reaction occurs in that the vehicle makes an unexpected sharp (although short) turn, which at a medium speeds (30-35 km/h) can be unpleasant on dry roads and risky on slippery surfaces, since the dry clutching sticks fast.

On the contrary, the clutch of the dual-radius steering grips quite smoothly. The selected graduations for the two fixed radii offer very good adaptability to road conditions and a very smooth, uninterrupted drive, practically approaching that of continuous steering.

b) The constructive design of the single-radius steering also suffers from the following shortcomings:

1. Assembly and adjustment is unusually difficult and will most certainly give the front-line troops many problems.
2. The planned notch in the end position of the steering lever cannot be adequately felt. A fixed stop with a releasable lock is needed in order to switch over to track brakes from the radius clutch steering at any time without a time lag. When intending to engage the radius, it often occurs that the point for engaging the fixed radius clutch is skipped. If the steering lever is returned to its initial position the time loss is then so great that the intended curve can no longer be driven. The vehicle must be brought to a stop.
3. The clutch slipping needed for fine steering isn't possible, since the clutch is not designed for meeting such demands.
4. Reliable use of the steering is not always possible, as the performance of the oil pumps is apparently too dependent upon the r.p.m., so that the oil pressure is inadequate when the engine is turning over slowly. In critical situations, such as when the driver unexpectedly discovers he is not able to take the curve, he instinctively eases up on the gas and declutches. He can, however, steer the vehicle even less adequately due to the insufficient oil pressure. Therefore it must be required that the oil pumps deliver the necessary pressure for steering throughout the entire range of practical r.p.m.
5. The steering levers are very uncomfortably arranged. Their positioning does not permit the full power of the steering to be utilized when pulled back. With an increase in the pull on the steering levers power potential drops off instead of increases.
6. The single-radius steering mechanism requires follow-on adjustment to account for the wear placed on the support brakes, clutch and track brakes, which must be executed with care and which is impossible for the average driver and difficult for the majority of the tank maintenance crews.

7. Based on previous data, it is more expensive to produce than the dual-radius steering mechanism.

On the other hand, the dual-radius steering unit used in the Tiger offers the following advantages in addition to the aforementioned steering qualities:
1. A unique relatively simple assembly outside of the vehicle, with no adjustments necessary inside the vehicle. Meaning a short installation and removal time.
2. No follow-on adjustment necessary in the vehicle, requiring minimal maintenance.
3. Very practical utilization of space and spatial proportioning.

signed, Esser

Appendix 10

Panther mockups in use*

The creation of tank mockups is a means to deceive the enemy regarding our strength and intentions and cause him to take incorrect measures. The deception practice can serve the following purposes:
— Simulating our own tank concentration points in order to tie down enemy forces.
— disguising the breakup of our own tank concentration points.
— simulation of operational measures through rail transport of trains carrying mockups.
— In defensive operations by simulating counterattacks.
— Drawing off enemy aircraft and artillery fire from true targets.
— Deceiving enemy air and ground reconnaissance.

When utilizing mockups, the normal tank tactics of conventional units should borne in mind as far as is practical, in order to portray their operations in as realistic a manner as possible. This can extend to using containers of old oil to simulate hits from enemy fire, while the acoustic portrayal of tank sounds is accomplished by recording and loudspeakers.

Mass production of a suitable mobile Panther tank mockup requires:
— 3 carpenters and 3 assistants. Construction time 40 hours, work hours 6 x 40 = 240 hours.

The framework for the mockup is built using rough-sanded boards held together by nails. In order to produce a Panther mockup 2 cubic meters of cut pine wood is required. The base color of the mockup was yellow. The road wheels were outlined in black paint.

* Tank mockups were planned for the Panzer IV, the Panther and the Sturmgeschütz.

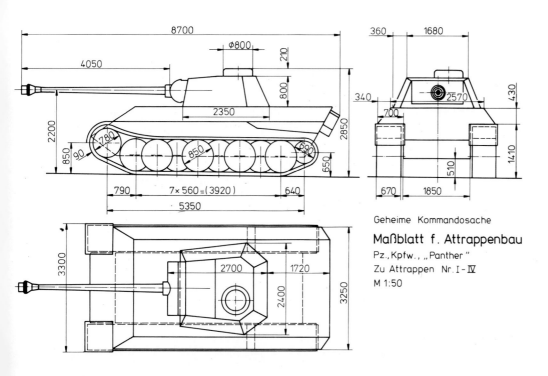

Appendix 11

Armor angles and thicknesses

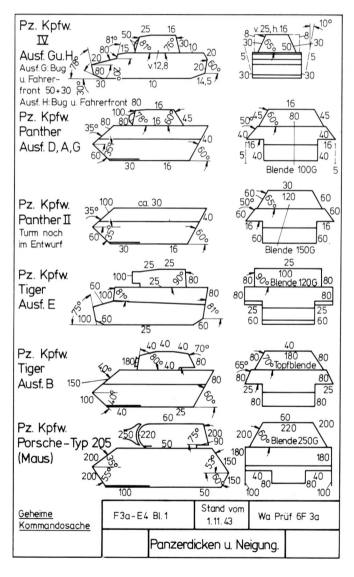
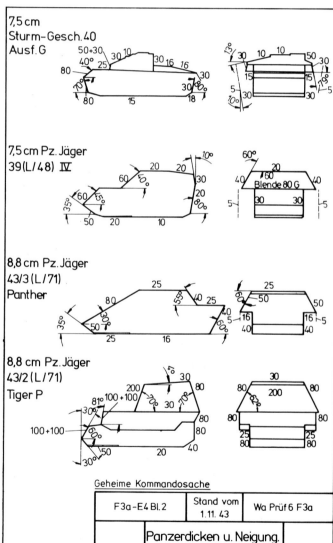

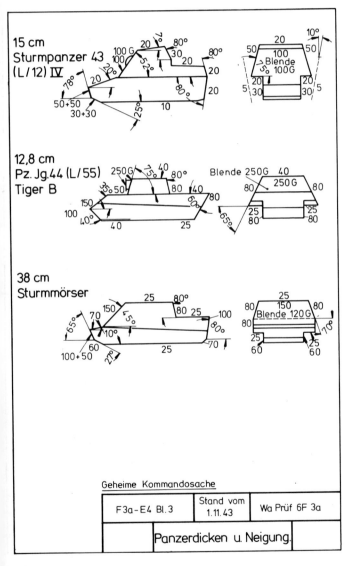
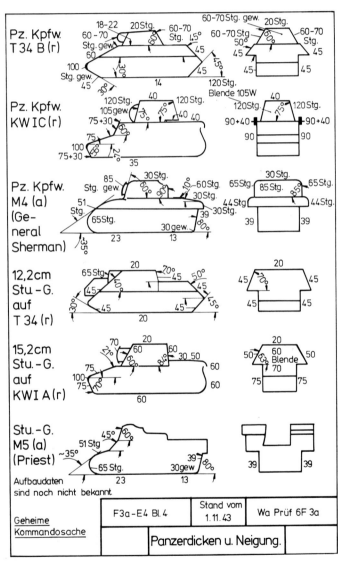

Only when using a 90 mm round was the Panther's glacis plate penetrated.

Left: shell damage to a Panther during firing tests conducted at the Aberdeen Proving Grounds in the USA.

Hits can be seen on the left turret wall. The heat developed on impact has caused the interior paint to burn off.

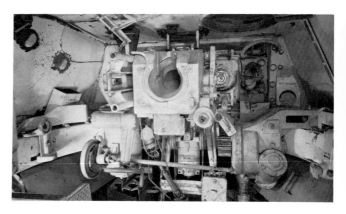

Appendix 11a

Results of shell penetration testing against armor

Inadequate material quality has caused the turret side plate to break — without achieving penetration of the armor.

Appendix 12

Panther production figures

Month/Year	Maschinenfabrik Augsburg-Nürnberg	Daimler-Benz	Maschinenfabrik Niedersachsen-Hannover	Henschel	WaA acceptance
January 1943	4	–	–	–	–
February 1943	11	6	1	–	–
March 1943	25	14	11	10	–
April 1943	–	90	39	26	–
May 1943	68	60	41	25	324
June 1943	43	40	36	25	160
July 1943	58	65	48	19	202
August 1943	38	26	39	15	120
September 1943	53	70	45	10	197
October 1943	104	90	50	–*	257
November 1943	76	71	75	–	209
December 1943	114	82	60	–	299
	549	614	445	130	1768
January 1944	105	90	75	–	219
February 1944	106	70	90	–	256
March 1944	96	85	90	–	270
April 1944	105	105	100	–	311
May 1944	125	110	111	–	345
June 1944	130	120	120	–	370
July 1944	135	125	119	–	380
August 1944	155	43	131	–	350
September 1944	140	80	120	–	335
October 1944	78	100	96	–	278
November 1944	103	110	100	–	318
December 1944	100	120	80	–	285
	1378	1158	1232	–	3717
January 1945	20	100	80	–	211
February 1945	22	70	81	–	130
March 1945	8	40	–	–	102
April 1945	20	–	–	–	–
	70	210	161	–	443
Total	2042	1982	1838	130	5928

* In the time frame from September 1943 to January 1944 an additional manufacturer (Demag?) produced a total of 50 Panther tanks. This brought the entire number of Panzerkampfwagen Panther produced to 6042.

Appendix 13

Compilation of engagement effects against the most significant enemy tanks by the German Pz.Kpfw. "Panther" (as of 30 May 1944)

The Panther can successfully engage the following listed vehicles:

Soviet-Russian T-34
front — to 800 m
sides — to 2800 m
rear — to 2800 m

Soviet-Russian KV 1 C
front — to 600 m
sides — to 2000 m
rear — to 2000 m

British Mark IV Churchill III
front — to 2000 m
sides — to 2000 m
rear — to 2000 m

American "General Sherman"
front — to 1000 m
sides — to 2800 m
rear — to 2800 m

Soviet-Russian Joseph Stalin with 122mm gun
front — to 600 m
sides — to 2000 m
rear — to 2000 m

The following enemy tanks pose a threat to the Panther at the following distances:

Soviet-Russian T-34 with 7.62 cm L/41.5 gun
front — to 500 m
sides — to 1500 m
rear — to 1500 m

Soviet-Russian T-34 with 7.62 cm L/30.5 gun
front — to 400 m
sides — to 1200 m
rear — to 1200 m

Soviet-Russian KV 1 C with 7.62 cm L/41.5 gun
front — to 500 m
sides — to 1500 m
rear — to 1500 m

Soviet-Russian KV 1 C with 7.62 cm L/30.5 gun
front — to 400 m
sides — to 1200 m
rear — to 1200 m

British Mark IV Churchill III with 5.7 cm L/45 gun
front — cannot penetrate
sides — to 500 m
rear — to 500 m

American "General Sherman" with 75mm L/40 gun
front — to 500 m
sides — to 1500 m
rear — to 1500 m

Soviet-Russian Joseph Stalin 122
Information regarding penetration characteristics cannot be given as the gun performance data is not yet available.

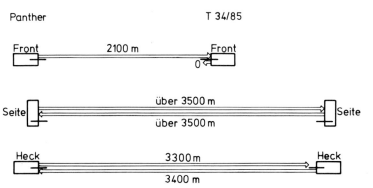

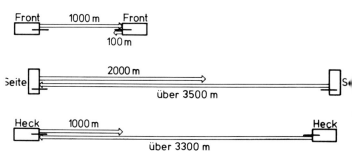

Appendix 14

Penetration effectiveness of the Panther gun

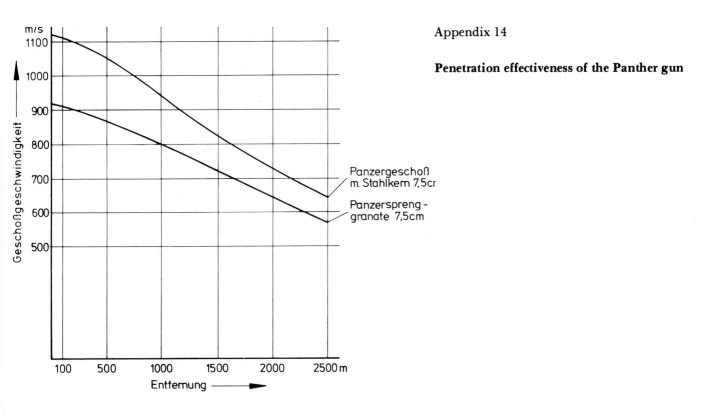

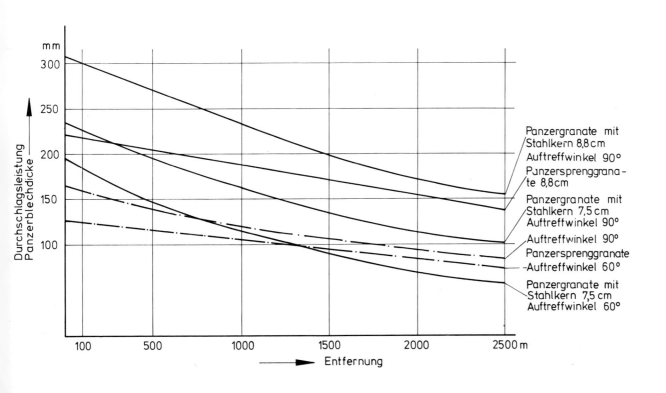

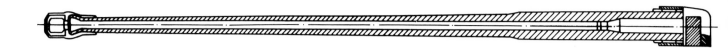

Rohr 7,5 cm Kampfwagenkanone 42

Appendix 15

Organization of a Panther tank maintenance unit

Vehicle composition of a Panther Panzerwerkstattzug (tank maintenance platoon)

1 350 ccm motorcycle
1 Funk-Kraftfahrzeug 15 radio vehicle
12 all-terrain 4.5 t spare parts trucks
2 Kfz 42 battery/communications workshop vehicles
2 4.5 t supply trucks
5 18 t SdKfz 9 half-track heavy prime movers
1 SdKfz 9/1 half-track with traversing crane (6 ton lift)
6 light all-terrain Personenkraftwagen
4 all-terrain Lastkraftwagen 4.5 t (with medium motorized maintenance platoon)
2 3 t all-terrain trucks for armorer-artificer
2 3 t supply trucks
1 medium bus
1 rotary crane truck (3 ton lift)
1 SdKfz 9/2 half-track with rotary crane (10 ton lift)
Trailers:
1 for communications shop
1 welding equipment
1 16 ton gantry crane
1 heavy Maschinensatz A generator
1 Sammlerladegerät D battery charger
1 for tent

Vehicle composition of a Panther I-Staffel
(Instandsetzungs-Staffel, or maintenance and repair echelon)

1 light personnel vehicle
1 SdKfz 3 Maultier outfitted for I-Trupp
1 all-terrain 3 t open bed spare parts truck
1 all-terrain 3 t maintenance truck for weapons and radio communication
1 all-terrain 4.5 t fuel truck
1 SdKfz 10 1 t light half-track
1 rotary crane truck (3 ton lift)
4 light all-terrain personnel vehicles
2 all-terrain 3 t truck
3 all-terrain 4.5 t spare parts truck, open bed
1 4.5 t supply truck
3 SdKfz 9 18 t heavy half-track prime mover
1 SdKfz 9/1 heavy half-track with rotary crane (6 ton lift)
Trailers
1 heavy Maschinensatz A generator (15 KVA 220/380 V)
1 battery charger

Vehicle composition of a Panther tank company I-Gruppe
(Instandsetzungs-Gruppe, or maintenance and repair section)

3 light all-terrain personnel vehicles
1 all-terrain 3 t open bed truck with equipment for I-Gruppe
2 SdKfz 10 1 t light half-track
1 all-terrain 2 t truck, open bed, equipped for I-Trupp
1 all-terrain 4.5 t spare parts truck, open bed

Each Panzer battalion operating the Panther tank had a Panzerwerkstattzug (Panther), or tank maintenance platoon. In addition, the supply company of the battalion also contained a maintenance and repair echelon, which in turn provided the individual Panther companies with maintenance and repair sections.

Maintenance units were equipped with a Strabo gantry crane with a lifting capacity of 16 tons; it was used to lift the turret.

Rotary crane truck with a lifting capacity of 3 tons, based on a Büssing-NAG 4.5 ton chassis, Type 4500 A.

Appendix 16

Organization of a Panther battalion

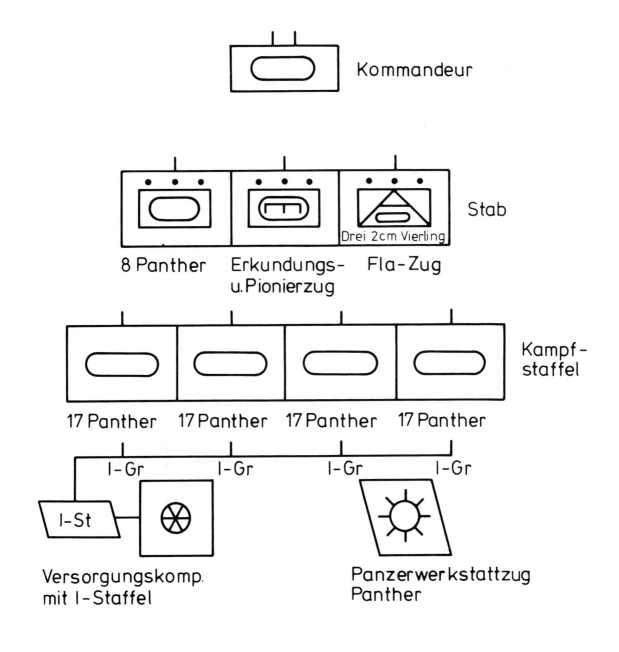

Appendix 17

Minutes from a meeting of the Panzer Commission

The final meeting of the Panzerkommission is included here, since it still exemplified the ideas, plans and projects which played an influenced the responsible offices of the time.

Re: Meeting of the Entwicklungskommission Panzer (Tank Developmental Commission) on 23 January 1945 in Berlin W 8, Pariser Platz 3, large conference hall Date: 23 January 1945

Attendance: all gentlemen on the included list, with the exception of:

Reichsminister Speer
Generaloberst Guderian
Hauptdienstleiter Saur
Dir. Dr. Blaicher
Dr. Rohland
Prof. Dr. Porsche
Dr. Maybach

1. Ammunition Carrier
Ardelt and Krupp are jointly developing a design for this ammunition carrier, which is to utilize the engine and transmission of the 38 D (AK 5 -80). Two examples are under development.

2. a) 75mm L/48 Recoilless Gun
Firing tests have been carried out using the 75mm gun and the recoilless gun has been deemed compatible for installation in the front plate of a self-propelled assault gun. The effect of the hard kickback upon the driver must still be observed. 5 examples stand ready.

b) Equipping of a Panther with an 88mm L/71 (6.20 m)
This study is being carried out by Daimler-Benz.

c) New Armament for Armored Vehicles
A new type of mortar which can deliver a rocket-propelled projectile to the enemy up to a distance of 2000 meters is currently under discussion. Oberst Geist, Schaede and Holzhäuer took part in this dialogue with considerable interest.

3. Flame-throwing Tanks
Up to this point, maximum ranges of no more than 60 to 70 meters were possible using pump-operated nozzles. A new type with highly-compressed nitrogen in on-board containers permits a range of 120 to 140 meters. Since these containers only are sufficient for 10 to 12 flame burst of 1 second each, a larger tank is considered necessary.

4. Status Report
General Thomale gave an insight into the tank situation. We have lost large amounts of spare parts in the East; heavy armored vehicles have also been lost. He therefore calls for:

1. conservation measures, delivery of spare parts and recovery vehicles.
2. As many components as possible should be manufactured within the constraints of the raw material shortage.

Accordingly, he welcomes the efforts in the area of the 38 D assault gun design. Stemming from this material shortage the tank has presently become a defensive anti-tank vehicle. This, however, is in no case due to a superheavy gun and thick armor plating. He relates how for the past two years countless thoughts and recommendations have been concerned with the so-called Panzerkleinzerstörer, the U-Boot of the land. Representatives of the Luftwaffe have already been to these meetings, desiring to introduce such vehicles into their paratrooper units.

Oberst Holzhäuer related that he has worked on approximately 20 designs since the spring of 1942, among which the building projects of BMW (Rutscher) and Weserhütte have become known. Since in the meantime nothing further has occurred, Büssing is to be contracted to build a two-man vehicle of 3 tons net weight with a 90 hp engine based on the Tatra 4 cylinder engine.

A compromise solution, making use of available components, is a proposal by Humboldt-Deutz. This vehicle weighs approximately 5 tons, meaning that it is not suitable for airborne operations. Five such vehicles are to be quickly built in the near future.

5. Panzerjäger 38 D

Herr Michaels of Alkett reported on planning for this vehicle. Increasing the engine torque from 42 to 78 mkg by using the Tatra diesel engine is expected to give the vehicle considerable maneuverability. The simple solution of equipping the 38 D with the new engine was apparently impossible. The BMM design of the 38(t) is not suitable for a rational mass production program due to the inordinately great amount of hand finishing required. The installation of the air-cooled diesel engine brought a series of new arrangement problems. The previous weight of 14.5 tons climbed to a combat weight of 16.7 tons, including 62 rounds of ammunition (1000 kg). The gearbox had double the torque compared to the Wilson gearbox, yet only required a 15% weight increase. Fuel consumption of 76 liters per 100 km on smooth terrain gives a range of 510 km with a fuel capacity of 390 liters. With the Tiger only about 100 km can be attained. He offered assurances that the design is the best for series production of tanks. The variants possible with this vehicle are:
1. Jagdpanzer with L/48 gun
2. Reconnaissance vehicle
3. Jagdpanzer with L/70 gun
4. Armored reconnaissance tank with 20mm gun and gun shield
5. Armored reconnaissance tank with superstructure 75mm gun
6. 120mm mortar
7. Kugelblitz

and two additional vehicles for which no further information is given.

Another new running gear supporting a 20 ton vehicle weight has been developed for this vehicle by Alkett using volute springs instead of torsion bars.

6. Assault Mortars on the Hummel or Tiger

Attempts are being made to install these mortars not in the Tiger, but in lighter vehicles.

7. Developmental Vehicles

Oberst Holzhäuer reports that the E-10 and E-25 with rear-mounted electric transmission should be continued if possible. The E-10 has already been stricken from development and supplanted by the 38 D.

Oberst Holzhäuer is in agreement with the continued development of the E-25 tank with a 400 hp MM engine or a 350 hp air-cooled Argus engine.

Oberst Schaede inquired as to the state of testing. The hulls are in Kattowitz and, on recommendation of Holzhäuer, are there to be picked up by a specialized tank transporter. Herr von Heydekampf is of the opinion that the E-25 could fill a void at some time in the future in the weight class between the 38 D and the Panther. The running gear is a plate spring design.

8. Tank Engines

Herr von Heydekampf indicates that the primary goal is greater horsepower and engine torque, then diesel power and air cooling, whereby naturally his chief concern is to move the 75 tons of the King Tiger. Considering that the MM engine with a choke r.p.m. of 2600 1/min only delivers barely 540 hp, this means that only 7.3 hp per ton is available. Herr von Heydekampf expects the quickest solution with the MM HL 234 engine with fuel injection in place of a carburetor. General Thomale expressly requested that it not be put into series production as quickly as with the HL 230. The engine must have reached 100% maturity when it reaches the front-line troops.

Oberst Holzhäuer gives the status of work on the new diesel engines. The Führer directive calling for the construction of a diesel motor with air cooling is still in effect. The engine from Simmering-Pauker-Porsche has already been blocked and has been running for some time. The cooling fan, developed by Professor Kamm, is not yet adequate. Within a year, roughly in nine months, the two-stroke Deutz engine is expected. This still utilized water cooling, but otherwise has enticing simplicity. In no case should this be suspended by the Panzerkommission. As opposed to the Simmering-Porsche 16 cylinder engine, it displays less weight and a particularly easy accessibility to all working parts. It was claimed that the Simmering engine requires 100% more material expenditure than the MM engine and is not interchangeable in the vehicle with the MM engine. Compared with the Panther II's expanded compartment and that of the

Panther I, the King Tiger's engine compartment is so narrow that the new engines cannot be fitted. Only the MM-HL 234 is possible; entirely new hulls will have to be developed for the new engines.

Herr von Heydekampf concludes and comes out in favor of the immediate introduction of the MM-HL 234. This has greater power, instead of 700-900 hp, more rpms and a favorable specific fuel consumption of 220 g per hp per hour. There is also less load on engine parts. Herr Ochel requests an early decision so that setup can get underway using all means possible. Maybach has finally moved to strengthen the piston rods and improve the cylinder head gasket. Herr von Heydekampf will give the go-ahead in approximately four weeks.

Herr Ochel reports that the production facilities for the Simmering-Porsche engine are being established at Steyr under the leadership of Direktor Meindl and that an initial production run of 2000 pieces has already begun. He had wanted Steyr to set up for the eight cylinder Tatra diesel engine for the 38(t). Steyr, however, had refused, since there is no foreseeable way to initiate the production of two engines. Series production of the Simmering engine should begin in June/July 1945. Ochel points out that the engine has not yet undergone driving trials and that this initiation of series production is somewhat premature. Herr von Heydekampf desires to interject into the Panzerkommission proceedings and, if possible, stop the production immediately. In this effort Heydekampf is supported by Holzhäuer. Holzhäuer wanted the eight-cylinder Tatra engines for the 38(t), the 8 ton half-track prime mover and the SWS, i.e. the AK 5-80 transmission will have to be used in these vehicles as well.

In order to provide an alternative to the Tatra diesel engine, Maybach has proposed the HL 64 with fuel injection with roughly 270 hp in place of the 210 hp of the Tatra engine. Moreover, Dr. Maybach has directed a letter to the Panzerkommission in which he expresses his doubt that the gasoline supply situation is so problematic that it is absolutely essential to use diesel engines.

9. Final Drive
From the front there continues to be serious complaints regarding final drive breakdowns in all vehicle types. Approximately 200 breakdowns have been reported with the 38(t). Prior to the 1945 eastern offensive there have been 500 defective final drives in the Panzer IV. From the Panther 370 and from the Tiger roughly 100. General Thomale explained that in such circumstances an orderly utilization of tanks is simply impossible. The troops lose their confidence and, in some situations, abandon the whole vehicle just because of this problem. He requests an increase in efforts for the final drive, since only this way can the problem be laid to rest. With the previously intense criticism of the engine and the final drive continually playing such a roll, it is welcome news to learn that the gearbox generally enjoys a good reputation. Direktor Wiebicke claims that the Heerestechnisches Büro of the Waffenamt had for its part rejected the sun-and-planet final drive and demanded the spur wheel reduction drive. This claim led to a confrontation between Oberst Holzhäuer and Oberbaurat Knönagel. Oberingenier Wiebicke clarified that for the past one-and-a-half years there has been ongoing discussion regarding the introduction of the planetary gearing but as of yet nothing significant has been accomplished. Whereas during this entire time attempts have been made to improve the final drive, with only minimal improvement being noted. It must, however, be kept in mind that MAN, as the responsible manufacturing firm, cannot now hold other companies responsible. MAN has availed itself of all offices which have the prospect of providing a way out of these difficulties with the final reduction drive.

10. Air Filters for Tanks
There is nothing to report on this matter.

11. Milling Tolerances
A reduction in tolerance could not be agreed upon among the companies supplying steel. On the contrary, there is the prospect that the tolerance will become less refined.

12. Engine and Gearbox
a) New stop valves and gaskets are being used in the MM engine.

b) Oberst Holzhäuer reports on the solid, liquid and gaseous alternate fuels. Up to now, the only acceptable alternate fuel has unfortunately been gasoline. Road testing vehicles using gas generators is rejected by him. There is a danger of seizing and because of that the long-term damage to the engine during startup is great. A short production run of gas generators has been completed, although their use for road testing has not been resolved. It appears that the MM company fears for its engines and therefore has pushed its refusal stand through the bureau.

There were attempts to reduce the road test time by using gasoline, but the lower acceptable limits have already not been met.

c) Oberst von Wilcke posed the question to the Kommission as to whether the AK 5-80 gearbox for the Panzer II and IV and the AK 7-200 in the Tiger II must be included in the planning for replacement supplies in the immediate future. He certainly did not want to bring up the issue of OLVAR-AK gearboxes before the Panzerkommission again, but he must have clarification on the matter of planning. Herr von Heydekampf explained that the constant mesh type has proven itself and from his viewpoint has nothing against introduction into the vehicles cited. On the other hand, Herr von Heydekampf considers the installation in the Tiger undesirable, as the changes needed inside the vehicle for installation would be rather comprehensive. Oberst Holzhäuer considers it necessary to quote an opinion expressed by Herr von Westerman, calling the test installation of the AK 7-200 in the Tiger an "emergency remedy." In addition, the 75 tons vis-a-vis the 48 tons of the Panther should also be considered. I explained that we ran two AK 7-200 gearboxes in a King Tiger in Kummersdorf and their reliability need not be questioned

Closing Remark:

Herr von Heydekampf closed the meeting by expressing thanks to those who were present despite the travel difficulties, and in particular thanked those who did not show up — since the discussion was thereby shortened without any adverse effects.

Nothing was mentioned on the military situation.

Although Reichsminister Speer, General Guderian and Hauptdienstleiter Saur were not present, even though the latter had stated he would be, this is nevertheless to be considered as a good sign.

signed, Maier
31 January 1945
TBE Mai/He.

List of participants at the meeting of the Entwicklungskommission Panzer (The Reichsministerium für Rüstung und Kriegsproduktion on 23 January 1945, Berlin W8, Pariser Platz 3

Oberst Geist: Ministerium für Rüstung und Kriegsproduktion
Oberst Schaede: Ministerium für Rüstung und Kriegsproduktion
Generalmajor Thomale: Generalinspekteur der Panzertruppe
Generalltn. John: Heereswaffenamt (Amtsgruppe für Entwicklung und Prüfung)
Generalmajor Stammbach: Heereswaffenamt/Wa J Rü (Amtsgruppe für industrielle Rüstung und Beschaffung)
Generalltn. Beisswänger: Allgemeines Heeresamt/In 4 (Artillerie)
Oberst Proff: Allgemeines Heeresamt
Oberst von Wilcke: Heereswaffenamt/WuG 6 (Panzerbeschaffung)
Oberst Holzhäuer: Heereswaffenamt/Prüf 6 (Entwicklung Motorisierung und Panzer)
SS-Gruppenf. Dr. Schwab: Waffenamt der Waffen-SS
Min.-Rat Röver: Heereswaffenamt/Wa Chef Ing 4 (Fertigung)
Min.-Rat Dr. Leinweber: Heereswaffenamt/Heerestechnisches Büro
Ing. Krömer: Ministerium für Rüstung und Kriegsproduktion (Tiger und Panther)
Oberst Dr. Körbler: Allgemeines Heeresamt/In 6 (Inspektion der Panzertruppe)
Dr. Stieler von Heydekampf: Henschel & Sohn, Kassel (Vorsitzender der Entwicklungskommission Panzer)
Dir. Dorn: Krupp, Essen
Dir. Wiebicke: MAN, Nuremberg
Dir. Wunderlich: Daimler-Benz, Berlin-Marienfelde
Dipl.-Ing. Michaels: Alkett, Berlin-Borsigwalde
Ob.Ing. Maier: ZF., Friedrichshafen a.B. (Getriebering)
Prof. Dr. Ing. Osenberg: Technische Hochschule Hannover/Fertigung
Prof. Dr. Benz: Amt für Bodenforschung/Materialien
Dir. Ochel: Orenstein & Koppel, Berlin (Motorenring)
Dir. Welge: Brandenburger Eisenwerke, Brandenburg (Panzerstahl)
Oberst Zierhold: Allgemeines Heeresamt/In 4 (Artillerie)
Oberst Wendscher: Allgemeines Heeresamt/In 4 (Artillerie)
Oberst Wöhlermann: Heereswaffenamt/WaPrüf 4 (Waffenentwicklung)
Oberstltn. Otto: Heereswaffenamt/WaPrüf 1 BuM H.-Versuchsstelle

Appendix 18

Internal combustion gasoline or diesel engines for the tank

Prof. Dr.-Ing. R. Eberan von Eberhorst, Institut für Kraftfahrwesen der Technischen Hochschule Dresden*
With the development of the diesel engine as a high-speed motor, it has grown into a serious contender for the Otto or gasoline engine, and as a result of its greater economy despite a complicated design has found increasing utilization particularly for heavy truck transports. Nevertheless, its use in the tank remains contested.

Here in this synopsis the requirements expected of a tank engine are presented and, based on the state of technology, to what extent these requirements are better met by one or the other engine design. Its worthy to note that of the 35 enemy tank types known to us, only 7 are equipped with diesel motors (as shown in diagram 1), the remainder use gasoline engines for vehicles or special-designed engines. The diesel and internal combustion is represented in a 4:16 ratio.

The medium to heavy tank requires a power capability of 300-700 hp for attaining the requisite high terrain mobility, necessitating a high performance motor, the performance of which lies between the standard vehicle engine and an aircraft motor. The changing tractional resistance of 2 to 80% of the vehicle weight encountered in cross-country operations requires a dependable power controllability, especially a high tractive power capability at low r.p.m., for which the expression "Büffel-Charakteristik" (lit. "buffalo characteristic") was coined. The engine qualities should make frequent gear changing superfluous. The motor should have a large speed range. Cooling must be adequate even under extreme climatic conditions and operate independent of the vehicle's speed. Given the variable outside temperatures, the cooling must be controllable in a simple manner. The engine should be particularly economical in order to give the tank the greatest radius of action possible. The highest operational reliability, straightforward care and maintenance, good accessibility to components to be worked on or changed out — all these are to be expected. The life of those parts subject to wear should be, at the very least, 5000 km driving distance. Finally, the last and most important prerequisite is the minimum possible space requirement for the entire drive unit, including the engine cooling system, fuel tanks and transmission.

Among our enemies, only the Russian V 2 diesel engine is considered to be a special development for tanks. Otherwise, proven available engine design models were utilized, among these being aircraft engines. With regards to the gasoline engine used by us in our tanks we cannot expect a conclusive answer to the question "gasoline or diesel engine" to be gleaned from the design of the enemy tank engines. It is far more important to weigh the features of both engine operations and designs against each other and compare as many proven engines as possible.

We begin then here with output efficiency N and maximum pressure ppp of the complete engine unit, which in the diesel motor causes a higher engine loading. Thanks to its higher compression ratio and the higher air coefficient, the thermal efficiency is considerably greater than with the gasoline engine. The efficiency certainly drops when limiting the maximum pressure, but the engine can then be constructed more easily.

Example	**Gasoline**	**Diesel**
compression	* = 6	* = 15
air quantity	* = 1.0	* = 2.0
maximum pressure	p_{max} = 55 at	p_{max} = 110 at
engine load	* = 100%	* = 200%
net efficiency	* = 0.38	* = 0.59
indicated horsepower	N_i = 100%	N_i = 78%
specific consumption	b = 100%	b = 85%

---

* Excerpt from a report given in the Reichsministerium für Rüstung und Kriegsproduktion, Berlin, on 26 April 1944.

According to the thermal balance of two aircraft engines, the diesel motor converts 39% of the excess heated air taken in into effective output, whereas the gasoline engine only utilizes 25%. The excess warm air being drawn off by the cooling unit of the gasoline engine is more than double that of the diesel engine, which is important for the cooling system in terms of spatial, power and weight constraints. The energy lost in exhaust is roughly 70% higher at the same power output for gasoline engines.

The significance of the engine torque and performance as it relates to the driving characteristics of the tank is shown in Diagram 2. An increase in engine torque with r.p.m. dropping off ("Büffel-Charakteristik") offers better hard pulling than the flat engine torque characteristics of a high-performance engine. As an example, with equal maximum output an engine with 33% torque increase and 5-gear transmission has the same qualities as a motor with flat-running engine torque and a 9-gear transmission.

The fully-loaded engine torques of some gasoline and diesel vehicle engines (Diagram 3) show that the high speed gasoline engine regulated by a carburetor more closely approaches the feature of the "Büffel-Charakteristik." Its maximum engine torque increase of 40% (opposed to 27% with the diesel engine) is not by chance (a greater increase in engine torque for the diesel engine would be possible through a fuel injector with each power stroke of boosted capacity at decreasing r.p.m.).

The biggest advantage of a diesel engine is its lower fuel consumption. However, it is not sufficient to simply compare the lowest values of the specific consumption with each other. It is much more insightful to compare average values in a broader performance range (25-100% NNN) at the most common running r.p.m. (n=2/3 NNN) (Diagram 4).

The average consumption values, which can then be used as the basis for developmental prospects of both approaches, are taking into account the power expended in cooling)
for the (best) gasoline engine bm = 257 g/PSh
for the (best) diesel engine bm = 171 g/PSh
(100:66.5)

Diagram 5 shows the most important design and operating features of 126 gasoline and diesel engines for both indigenous and foreign vehicles (Kfz), tanks (Pz) and aircraft (FL). The bars in black represent the values for diesel motors, while the lined bars are for gasoline engines (lightly lined/crossed bars are best values). The figures in the lower right are the serial number of an engine performance overview chart.

The higher average pressures of aircraft engines stem from supercharging, fuel injection and the utilization of high octane fuels. In terms of piston speed the best vehicle and tank engines approach the values for the aircraft engines very closely, whereas the piston surface output, being a product of average pressure and piston speed, drops off considerably with vehicle engines when compared to aircraft motors. The effects of the commandment calling for a lightweight design, which is of primary concern for the aircraft engine, is manifest in this engine's excellent power-to-weight ratio (also indicated here as kilograms per metric horsepower (kg/hp or kg/PS)). The best tank engines are significantly superior to vehicle engines in the performance to weight category, because the design in question is more closely related to the aero engine.

When the rectangular block dimensioning ascribed to the engine in cubic meters is related to piston displacement this factor then gives a measure for space utilization. Using this concept, it is worth mentioning the outstanding space utilization factor of the Russian V 2 tank diesel engine.

The displacement qualities of vehicle and aircraft motors show that the gasoline engine is only marginally superior compared with the supercharged two-stroke diesel engines. The best value for a tank diesel engine is also a two-stroke engine. A further characteristic affecting space utilization is the power output-to-space ratio (in hp/m³) of the applied rectangular area (here the maximum performance figures given by the manufacturers were used!). As expected, the aero engines are significantly superior in terms of their performance-to-space compared with vehicle engines. The comparison of gasoline/diesel engines in the category of tanks is unreliable, as there is not a sufficient range of designs presently being used operationally. A more accurate performance-to-space comparison between gasoline and diesel motors is provided by the aero engine. The superiority of the gasoline engine in terms of performance-to-space must remain low for tanks, due to the fact that only low-OZ fuels are available.1*

The average specific consumption and its reciprocal value (PSh/kg) is approximately the same for the best vehicle and tank engines, but circa 31-33% better for a diesel engine.

The values for tank engines presently under development (DE) are compared with German and foreign engines (D+A) currently in service in Diagram 6. The performance values p_m, c_m and N/F of the gasoline engines are still below the current best values, with the power/weight (N/G) being above the average. In all categories the diesel engines under development are higher than the current designs. Power/weight (N/G), space utilization V_h/N_g, power/space N/V_g of the individual designs, however, vary considerably. But even here the performance-to-space ratio of the engine cannot in itself be used as a measurement for the spatial requirements of the drive unit, since air-cooled motors are being compared to water-cooled engines — the latter requiring almost as much space for the cooling system as for the engine itself. The external form of the entire engine including all components and auxiliary devices dictates the space requirements, whereas the radiator, exhaust feed and even the air filters can generally be fitted into the existing space. Due to its restrictive design, the engine's rectangular blocking is disproportionately large, leading to dead spaces which are hard to exploit.

Aside from the space utilization already established, two characteristic volumetric efficiency scales, where $V_r = V_g$, are outlined for spatial utilization of the design model in question. Silhouette area was compared with rectangular blocking, both in sectional and cross-sectional cuts. If the space $V_r = V_g$ does not accurately correspond to the actual design volume of the engine, it probably gives a more accurate measurement for spatial installation requirements than the rectangular blocking.

The tank engines, of which a few designs in use are compared in Diagram 7 with newer developments (or engines previously used for other purposes), show that both gasoline as well as diesel motors, when conforming to the tank hull's restricted dimensions, offer utilization of a relatively large piston displacement and relatively high performance in a confined space. A cross sectional view of the Panzerkampfwagen V is provided for illustrating the spatial relationship to the Motor 95 in scale. The best utilization of space is achieved by the V engine, which is superior to other designs in terms of accessibility.

The last diagram, number 8, shows the specific spatial and weight requirements of the entire drive unit including the transmission, cooling unit and the fuel tanks, the size of which is dependent on the proscribed combat radius.

The space requirements in l/hp increases with operating time and average fuel consumption, based on a fixed value set by design for the engine, the cooling unit and the transmission collectively.

In view of its space-conserving design, a gasoline engine of the same lightweight construction and with the same displacement as the Russian V 2 diesel could exceed the performance of the diesel engine for up to 6 hours operating time despite the greater spatial requirements for its cooling system. Over longer periods of operation the lower fuel consumption, expressed in the more gradual curve incline, works in favor of the diesel engine. Two diesel engine developments are clearly better than the V 2. In weight, no less a decisive factor, only one engine project is better than the V 2. Even the ideal gasoline engine can hardly exceed the performance of a diesel engine when the entire drive unit is compared. The values shown in Diagram 8 for the spatial and weight requirements are evident from the table.

Based on preceding models, efforts should be taken to incorporate the following design and operating features for the tank engine of the future:

1. Engine with large piston displacement in the most compact space, therefore cylinders arranged in V form.
2. Two-stroke operation due to greater power per unit of displacement
3. Diesel operation due to
a) less fuel consumption
b) smaller cooling system
c) larger r.p.m. range
d) power by both diesel and gasoline fuels
e) reduced danger of fire
f) complete immunity to inclined angles
g) radio interference suppression not necessary
4. Performance boost through supercharging

Constant operational readiness is just as critical as the number of tanks we send to the front. Accessibility to parts needing repair and prone to breakdowns cannot be determined on the drafting board; the front line has the final word. To relieve the front of this burden to a greater extent than before is the goal of every new development. The best tank motors and engines — primarily including those of our enemy — are taken, and repair/replacement times are determined (without removing the engine) for cylinder heads, pistons, valves, etc. Then clear maximum time limits are established that are tested and applied to models and prototypes; all this in order to spare the front-line troops losses in blood and materiel.

Appendix 18a

Comparison of Engines

Make	model	gasoline/diesel	stroke	liquid/air	type	bore(mm)	HUB(mm)	volume	rpm	performance (hp)	passenger/truck
5 Büssing-NAG	GV6	gasoline	4	liquid	6!	130	170	B13539	1600	174	
16 Maybach	HL 108	gasoline	4	liquid	12V	110	115	10835	2600	239	
25 Skoda	Rapid1500	gasoline	4	liquid	4!	72	96	1569	3500	42	
26 Skoda	Superb 3000	gasoline	4	liquid	6!	80	104	3137	3500	80	
27 Tatra	87	gasoline	4	air	8 V	75	84	2970	3600	70	
28 Tatra	92	gasoline	4	air	8 V	80	99	3990	2500	70	
47 Deutz	AF 8 M 114	diesel	4	liquid	8!	105	140	9680	2400	220	
48 Deutz	AF 6 M 517	diesel	4	liquid	6!	130	170	13500	1500	190	
50 Einheitsdiesel	HWA 538	diesel	4	liquid	8 V	105	130	9008	2400	130	
58 Skoda	254	diesel	4	liquid	4!	100	120	3768	2200	55	
59 Skoda	706	diesel	4	liquid	6!	110	150	8553	1800	107	
67 Saurer	CHDv	diesel	4	liquid	8 V	110	140	10640	2200	160	
68 Saurer	CKD	diesel	4	liquid	6!	95	125	5310	3000	100	
Tanks											
70 Praga	AC 1	gasoline	4	liquid	6!	110	136	7750	2800	180	
71 Praga	NR I	gasoline	4	liquid	8 V	125	175	14450	2400	300	
72 Skoda	T 15	gasoline	4	liquid	8 V	115	130	10802	2800	235	
73 Skoda	T 22	gasoline	4	liquid	8 V	130	140	14860	2200	263	
76 Maybach	HL 62	gasoline	4	liquid	6!	105	120	6235	2600	139	
77 Maybach	HL 120	gasoline	4	liquid	12 V	105	115	11952	3000	300	
78 Maybach	HL 230	gasoline	4	liquid	12 V	130	145	23100	3000	700	
79 USSR	V 2	diesel	4	liquid	12 V	150	183	38880	2000	600	
85 F.K.F.S.	(Kamm)	gasoline	4	air	2x12 v	135	140	48000	2050	1000	
86 Simmering(Porsche)	101/4	gasoline	4	air	10 V	115	145	15060	2600	345	
87 Daimler-Benz	MB 507	diesel	4	liquid	12 V	158	180	42300	2300	850	
88 Daimler-Benz	MB 517	diesel	4	liquid	12 V	162	180	44500	2400	1250	
89 Deutz	T8 M118	diesel	2	liquid	8 V	170	180	32300	2000	700	
90 Deutz	T8 X113	diesel	2	liquid	8*	130	130	13800	2500	400	
91 Deutz	Dz 710	diesel	2	liquid	16-.-	160	160	51500	2500	1500	
92 MAN/Argus	LD 220	diesel	4	air	16 H	135	165	37800	2200	700	
93 MAN	1038 GL	diesel	4	liquid	8 V	110	130	10000	3000	200	
94 Saurer	Z. 2-T	diesel	2	air	8 V	105	135	9352	2800	274	
95 Simmering	Sla. 16	diesel	4	air	16 X	135	160	36500	2000	720	

Make numbers serve to identify the engines shown in diagrams

Clarification of symbols:
! = inline engine   V = V engine   * = radial engine   H = H engine   X = X engine   -.- = Boxer engine

Diagram comparison/gasoline or diesel engine

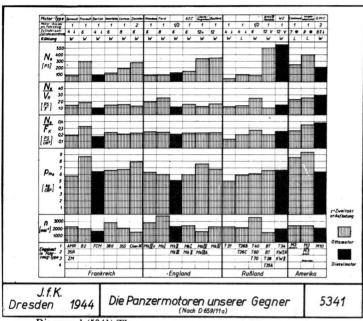

Diagram 1 (5341) The tank engines of the enemy

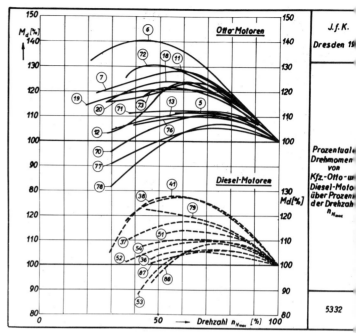

Diagram 3 (5332) Torque coefficient of truck gasoline and diesel engines

Diagram 2 (5337) Torque flow and number of gears

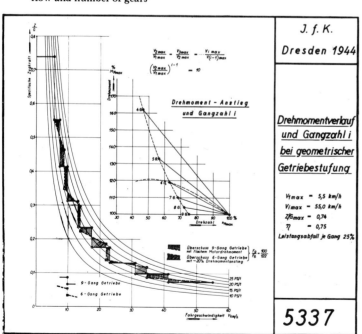

Diagram 4 (5333G) Specific fuel consumption of truck gasoline and diesel engines at 2/3 *

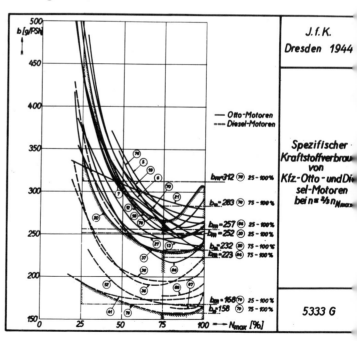

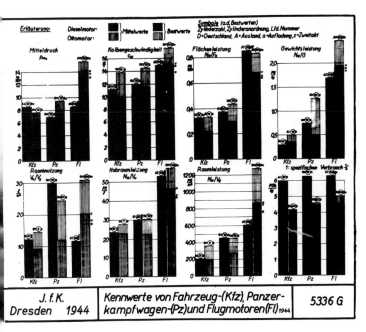

Diagram 5 (5336G) Characteristic values of truck (Kfz), tank (Pz) and aero (Fl) engines 1944

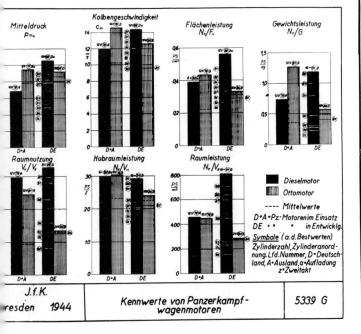

Diagram 6 (5339G) Characteristic values of tank engines

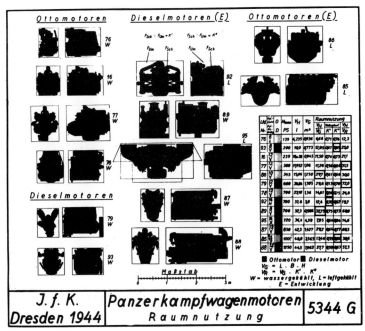

Diagram 7 (5344G) Space utilization of tank engines

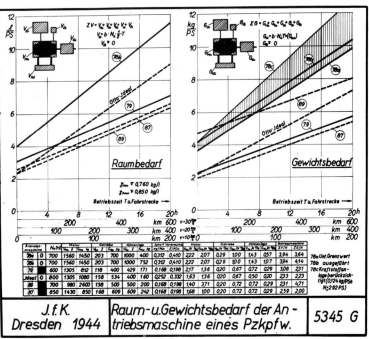

Diagram 8 (5345G) Space and weight requirements of the main drive engine of a tank

Appendix 19

Excerpts from a report by Oberst Dipl.-Ing. Willi Esser concerning newer tank models, given on 7 February 1945 before the Wehrtechnische Gemeinschaft des VDI in Berlin.

The tanks, tank destroyers and assault vehicles of the enemy are to be discussed within the framework of this report. The state of German development is illustrated using the Panther and Tiger tanks.

The significant features provided by the design and equipment which characterize such vehicles are:
the firepower,
armor protection,
mobility,
and means of communication.

The maximum speed of a large percentage of these vehicles is 40 to 50 km/h, the T-34 and the Panther approximately 55 km/h and the Cromwell and US M 5 65 km/h. As a scale for the available engine power, the concept of specific output was created, expressed in metric horsepower per metric ton (hp/t). The specific output only offers a rough clue for evaluation and only provides a scale for categorizing sizes based on engine performance to vehicle weight. The performance for most vehicles is between 10 and 15 hp/t; for those vehicles of a higher speed over 50 km/h this is between 15 and 20 hp/t.

For the Western Powers, particular significance is attached to the vehicles based on the Sherman chassis; for the Soviet Union this entails those vehicles based on the T-34 and KV chassis, with the follow-on development in the shape of the Joseph Stalin tank.

Engines (Diagrams D 19-21 Pz)
Whereas the Soviet Union utilizes a single engine type for five different vehicles, the number of engine types in the USA equals the number of tank designs. This stems from the fact that, initially, America used available motors which had been developed for other purposes and only later introduced an engine specifically constructed for installation in a tank. The diesel engine is utilized in a portion of US tanks and in all Russian tanks. Only the Wright-Whirlwind engine installed in some of the Sherman-series models, is air-cooled; the others are water-cooled. The Maybach HL 230 engine of the Panther and Tiger and the 8 cylinder Ford engine recently installed in the Sherman have been developed as tank engines. All remaining designs stem from other applications.

Two Cadillac automobile engines are installed in the American M 5 light tank. As of this time, four different engines are found to be in use in the Sherman-series vehicles captured; a fifth, the 9-cylinder Caterpillar diesel engine of 450 hp at 2000/1 min. is also known to be in use based on confiscated documents.

Diagram D 20 Pz shows the characteristics of the engines. The drive output to the fans and the power loss due to the air filters have not been factored out. The given performance of the majority of the engines lies between 400 and 500 horsepower, wherein it should be noted that the General Motors diesel in the Sherman III and M 10 is installed as a twin-engine and offers a total of 420 hp.

The Chrysler motor, created from five 6-cylinder engines linked together, generally gives the appearance of a makeshift design. Here five complete automobile engines with crankshaft are mounted together in radial fashion to form a single motor, the only modification being to the crankshaft housing. Each cylinder row has its own carburetor and ignition system. The Soviet Union makes use of the 12-cylinder diesel engine in the T-34 and variants and in the KV, its variants and its outgrowth — the Joseph Stalin. This engine delivers an output of 500 hp for the T-34 and 550 hp for the KV, the former having a maximum r.p.m. of 1800 1/min., the latter's maximum r.p.m. at 1900 1/min.

The Panther and Tiger have a 12-cylinder Maybach engine with a displacement of 23 liters. Particularly interesting is the comparison between the 8-cylinder Ford motor and the Maybach HL-230. Piston displacement rating, maximum constant r.p.m. and power-to-weight ratio are nearly the same for both engines. Diagram D 21 Pz shows the average characteristics for these engines as well as the two diesel engines. The maximum r.p.m. of the engines is assumed for each to be 100%. In order to better compare the course of the engine torque characteristics, the engine torque at a maximum r.p.m. of 100% was selected as a reference point. The sharply rising torque with decreasing r.p.m. for the V 2 and GM diesel engines provides good hard pulling ability at lower r.p.m., the so-called "Büffelcharakteristik." Even the 8-cylinder Ford motor is well designed with regards to this feature. The torque flow of the HL 230, being as it is of a different nature, is offset by the gear increments (7 gears in the Panther and 8 in the Tiger). It requires greater shifting frequency.

Diagram D 22 Pz gives an overview of the fuel consumption and range. During travel on paved roadways, the gasoline engine provides in liters per metric ton per 100 km between 6 and 9.5 liters, wherein the HL 230 in the Panther and Tiger offer the lowest rate; the highest is measured in the Cromwell with the Rolls-Royce engine and for the Sherman Vc with the 30-cylinder Chrysler engine.

For the vehicles with diesel engines, the fuel consumption is between 4 and roughly 6 liters per ton per 100 km. We find the lowest rate with the Russian vehicles and their familiar 12-cylinder diesel. The rate of just over 6 liters per metric ton per 100 km, high for a diesel engine, apply to vehicles of the Sherman series with the two stroke diesel engine and are in the same size class as the German tanks with gasoline engine. (Diagram D 114 Pz)

The fuel supply carried by the vehicles is between 11 and 22 liters per metric ton for the gasoline engines. Taking into account only the on-board fuel carried inside the vehicle, in the worst case (Sherman Vc) the range is 130 km; in the best case (T-34) it is 450 km. By including the fuel supply carried externally, the T-34 can cover a distance approaching 700 km. The great ranges which the Russian vehicles can cross are due to the low fuel consumption and relatively large fuel supply, which is not only carried in the engine compartment but also in the fighting compartment.

Gearboxes (Diagram D 23 Pz)
The development of gearboxes clearly shows the effort paid to alleviating the driver to a great extent of the burden of operating them. The Soviet Union has dropped the sliding mesh gearbox previously utilized in the T-34 and now only use those gearboxes which have clutch shifting in the upper gears.

All tanks in the Sherman series have 5-gear transmissions with clutch shifting and a synchronizing device, which only permits shifting once all shifting components have reached synchronization. The Panther tank has a manual 7-gear transmission with clutch shifting in all gears, ZF design, which is also called an "all-synchromesh" gearbox. All gears are synchronized, in a tapered arrangement. The Tiger is equipped with a semi-automatic 8-gear transmission which is operated hydraulically via a preselector. The well-known Maybach overdrive clutching serves to function as the coupling elements. Gearboxes with five gears are the most commonly used.

The maximum reduction, which determines the speed in low gear, is generally determined by the requirement for the greatest traction under rough terrain conditions, so that the slip limit is reached before the engine limits. This is therefore dependent upon the engine characteristics and the vehicle weight. Most transmissions are designed for a speed of 4-5 km in low gear. The overall ratio of the gearbox is determined by the speed in low gear and the maximum speed. For most of the vehicles of foreign make, this is somewhere between 9 and 10; the T-34 has a particularly low overall range of approximately 6, while it is a bit higher for the SU 85 assault gun (based on the T-34 chassis) due to the use of the 5-gear transmission. With approximately 13 for the Panther and 16 for the Tiger the overall range is the highest for German tanks.

Steering (Diagram D 24 Pz)
Clutch-and-brake steering are used in the tanks of the Soviet Union, Cletrac steering for those of the USA; the British and German Panther and Tiger tanks utilize controlled differential steering.

The clutch-and-brake steering for vehicles of the Soviet Union has multi-disk coupling, dry running brakes, and use gray cast steel as brake coating. (Diagram D 92 Pz)

The Cletrac steering in American tanks makes use of lubricated brakes with brake coating of woven asbestos imbedded with copper wire.

The single-radius steering for the Panther has, in addition to the primary drive coming off the gearbox, a source of steering power fed directly from the engine; this drives the sun gear wheel of a planetary gear system when the steering unit is engaged. When travelling in a straight direction, it is held in check by the support brake. When steering, this permits the drive sprocket on the affected side to be driven at a lower r.p.m. than when travelling straight. (Diagram D 94 Pz)

Since the primary drive power is transferred via the gearbox and the auxiliary drive comes directly from the engine, the steering radius is dependent upon the selected gear. Continued application of the steering lever activates support braking and a steering clutch. Steering brake is applied and work in the same fashion as with clutch-and-brake steering.

In addition to the primary drive, the double-radius steering mechanism of the Tiger also makes use of auxiliary drive power for steering, which is fed directly from the engine and works the sun gear wheels of the gear drive's planetary gearing. Steering clutches are closed during straight driving. The first step in steering is accomplished after releasing the steering clutch on the corresponding side. The clutch then is applied for the large turning radius and, with continued turning of the steering wheel, the clutch for the small turning radius — but only after the clutch for the large radius has been released beforehand. The drive sprocket is driven at a lower speed on the inner side of the curve and at a correspondingly faster rate on the curve's outer side. Since in this case as well the steering power comes directly from the engine, the turning radius is dependent upon the transmission's selected gear. (Diagram D 95 Pz)

Running Gear (Diagram D 25 Pz)

The design of running gear shows a development with the goal of providing the tank a more cushioned suspension. To this end torsion bars are utilized as suspension components in a portion of the Russian vehicles, in the new American Hellcat tank destroyer and in German tanks. The Cromwell and T-34 have coil spring suspensions; the M 5 light tank and tanks in the Sherman series (excluding variants) make use of volute spring suspension. The Cromwell, Hellcat, Panther and Tiger have hydraulic shock absorbers. For the running gear of the type used in the Sherman series, cushioning is accomplished through the gliding action between the running gear arms and the support mounts.

All tanks of the Soviet Union are devoid of shock absorbers.

With the exception of the Sherman series and the M 5 tank, road wheels are fixed to balance arms which are anchored in the vehicle hull. (Diagram D 115 Pz)

The average load per wheel width for rubberized road wheels fluctuates between 88 and 135 kg/cm.

The relative ground pressure is rather high for US tanks, at roughly 1.0 kg/cm^2, due to their narrow tracks, and at its lowest with the T-34 with 0.68 kg/cm^2. Follow-on developments of the T-34 and the Panther have a specific ground pressure between 0.8 and 0.9 kg/cm^2, the Cromwell at 1.0 kg/cm^2 and the Tiger a bit over 1.0 kg/cm^2.

Aside from the ground pressure, ground clearance is also a factor in determining cross-country capabilities. Efforts have generally been made to give the vehicles the greatest possible ground clearance. This is at its lowest with the T-34 at 380 mm and at its highest with the Panther at 540 mm.

Cast steel tracks are generally favored. The tracks for the heavy Russian KV and Joseph Stalin tanks are die-stamped. The Americans have gone from using tracks with flat rubber shoes and now use either non-skid rubber or steel tracks similar in form to the non-skid rubber, of welded or cast make, or even cast steel tracks with grip cleats.

The drive is at the rear for British and Russian vehicles; American and German drives are at the front.

Firepower

The firepower of a tank is a decisive factor in determining its combat value. It is subject to the weapon's effectiveness and the rate of fire; weapon effectiveness (Diagram D 89 Pz) being the
a) caliber
b) terminal velocity of the round
c) type of round
rate of fire being
a) method of acquiring, sighting and adjusting
b) vehicle position
c) ammunition (weight, dimensions, separate or cartridge ammunition)
d) weapon design (full or semi-automatic, or completely manually operated)
e) space constraints
f) workload distribution of the crew members
g) visibility after firing (muzzle smoke, dust)

The Soviet Union, with its 122mm gun in the Joseph Stalin, stands at the pinnacle in terms of gun caliber. On the other hand, the guns of the German Panther and Tiger tanks have a very long barrel length and, as we will see later, a superior penetrating power compared with all enemy tanks.

Only those tank guns with a caliber length of 40 and greater are considered good. The muzzle velocity of the rounds fired from these guns lies between 650 and 900 m/s.

The trend in development is in the direction of high initial velocities in order to provide greater penetrating power.

During the course of development the barrel length of the gun, measured from the total vehicle length, has grown considerably (Diagram D 28 Pz). With the increase in gun length the portion within the fighting compartment has also grown significantly and now requires a sizable amount of space inside the turret, which is increased even further through the use of longer ammunition.

In the Tiger the length from the trunnion pin to the baseplate is already greater than 1.60 meters. The cartridge it fires is 1.10 meters long (Diagram D 20 Pz). In order to be able to load, this results in a dimension of over 2.70 meters (calculated from the trunnion pin).

It is obvious that, with such measurements and with the diametrically opposed restrictions brought about by the load dimensions necessary for rail transport, the developmental trend is striving for limits, which cannot be overcome without revolutionary change. When using cartridge-type ammunition, the management of ammunition weight within the confined fighting compartment of a tank has practically reached its permissible limits with the 88mm caliber weighing roughly 23 kg. An increase in caliber, as has occurred with the Joseph Stalin, of necessity demands separate-loading ammunition; for the 122mm caliber the powder bag weighs approx. 14 kg and the shell weighs 25 kg.

With the limited carrying capacity, ammunition must be conserved. Therefore, when no specialized apparatae are utilized, the main gun must be fired from a halt. In several models of the Sherman tank, the Americans have introduced stabilization for the gun and, with it, for the gunner's optics as well; this system keeps the gun in the position set by the gunner over lightly broken terrain. With it, the gunner is given the ability of good terrain observation while the vehicle is moving, since the field of view remains steady. When the tank comes to a stop the gunner is ready to fire in the shortest time conceivably possible. The machine gun can be fired with accuracy when the vehicle is moving. At short ranges and against larger targets the main gun can also be fired with a relative degree of accuracy. The arrangement of the optics for the gunner has a disadvantage in this situation in that the viewport with head padding moves in conjunction with the relative movement between the gun and vehicle, meaning that the gunner is constantly having to move his head up and down.

The Soviet Union had earlier made limited use of a stabilized field-of-view for the optics. A prism in the optics was held steady in relation to the vehicle's movement by a gyro-stabilizer. In and of itself, such a method of stabilization does not permit an improvement in firing accuracy while on the move, but it does allow the gunner to maintain a good constant view of the terrain, acquiring and keeping the target in view. In this sense, a significant improvement in the rate of fire is achieved. (Diagram D 17 Pz)

For traversing the turret the Russians utilize an electric motor. In other countries hydraulic drives are used which work based on the displacement principle. These drives permit a very fine adjustment upon demand and a rapid slewing of the turret.

In the US tanks the hydraulic pump for the traverse mechanism is driven electrically. For the Panther and Tiger we use an adjustable constant hydrostatic drive of the same type as found in machinery construction. The drive is accomplished from the vehicle engine. The following times are established for a complete turret

revolution:
10 seconds in the T-34
30-35 seconds in the KV
35-40 seconds in the British tanks
15 seconds in the tanks of the USA
30-35 seconds in the Panther and Tiger

Armor Protection

In order to counter the increased penetrating power being achieved by newer weapons, there is a marked effort to provide the tank with strengthened armor protection.

The turret and forward half of the Joseph Stalin's hull are made of cast steel. Protection against frontal shots is offered by a refined shape. The body tapers to a point at the front in order to present the least possible amount of vertical surface area.

For the T-34 and the KV the general shape is the same as originally introduced. With the change in armament to 85mm the turret was modified on both and produced in cast steel.

In order to offer protection against air attack there has been an reinforcement of deck armor (Sherman IV, KV 85, JS 122, Tiger). The bottom armor is increased in the forward part (roughly a third to a half) against the effects of mines. The two data columns for the bottom plate thicknesses of the Cromwell, Sherman, Panther and Tiger in Diagram D 12g Pz indicate this strengthening.

With the exception of the cooling air intake and exhaust, the arrangement of the cooling system for the engine causes problems in the confined space and due to the requirement for good protection against small arms fire and shell damage.

This presents an overview of the current state of tank development. We have seen enemy and German tanks compared and do not need to shy away from this comparison.

The German inventive spirit in the best sense of the word, German thoroughness and creativity have presented a superior weapon into the hands of the German soldier in spite of every obstacle. On the engineering side of the house we soldiers express our thanks to the designers which, through secluded untiring labor, created these works. We are made aware of the tremendous developments in tank design nowadays when we once again quickly review the most significant examples that have made their appearance in this war: the French Char 2 C, which with its 72 tons caused such consternation prior to the war in the world press and among the public, but was never put into combat

the French B2 tank
the British Mark IV infantry tank (Churchill)
the US General Lee tank
the Russian T 35 tank, the great disguise for the T-34; at that time the best tank in the world
the KV II with the heavy gun (150mm caliber)
and the tanks of today:
the US Sherman IV tank
the Russian T-34/85 and Joseph Stalin
and the German tanks:
the Panther, which a year ago was recognized by the Russians as the best tank in the world,
and
the Königstiger, which was acknowledged by the Western Powers as, and in our opinion is today, the world's best tank.

Deutschland

 Pz.Kpfw. II (deutsch) 8,8t
 Pz.Kpfw. III (deutsch) 21,5t
 Panther D (L 71) 44,8t
 Tiger B (L 71) 68t

England–U.S.A.

 Mk VI C 5t
 M 3 „General Lee" 28 t
 M 4 „Sherman" 30 t
 M 1 „Dreadnought" 60 t

Rußland

 T 70 9,2t
 T 34 26t
 T 34-85 30t
 Josef Stalin 122 45t

Verskraft 1945 — Panzerkampfwagen Vergleich — **D 1 Pz**

Deutschland

 Stu.Gesch. III 24,5t
 Jagdpanzer 38 15,5t
 Jagdpanther 46t
 Jagdtiger 75t

England–U.S.A.

 7,62 cm Pz.Jäger M 10 30t

Rußland

 8,5 cm Stu.Gesch. T 34 30t
 12,2 cm Stu.Gesch. 30t
 15,2 cm Stu.Gesch. KW 43t

Verskraft 1945 — Jagdpanzer u. Sturmgeschütze Vergleich — **D 2 Pz**

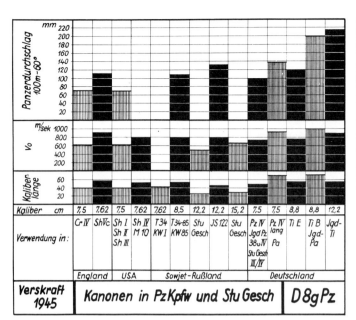
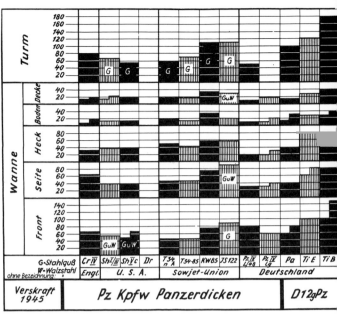
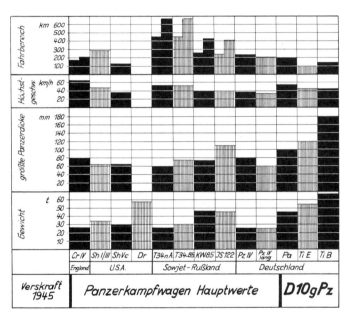
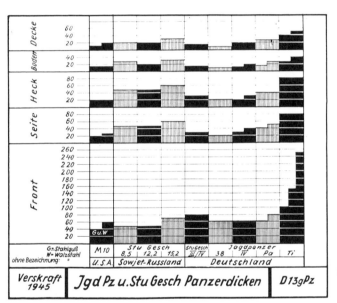
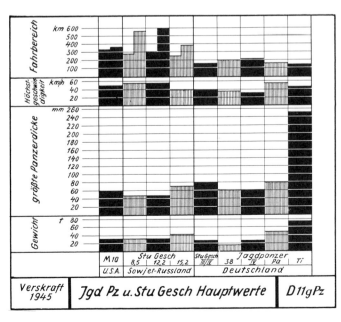
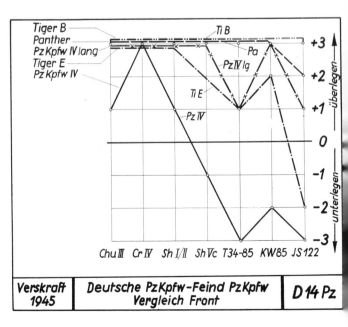

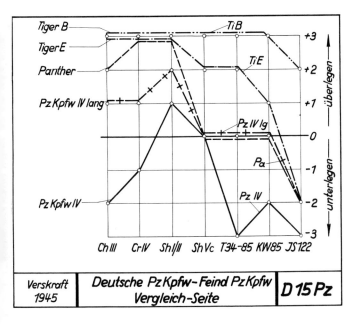
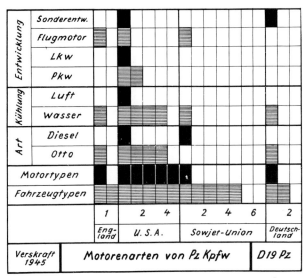
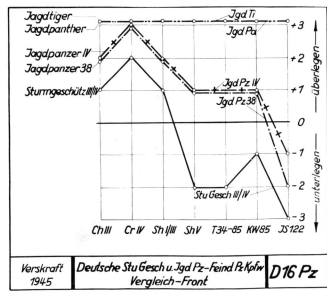
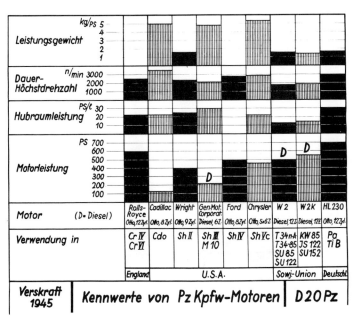
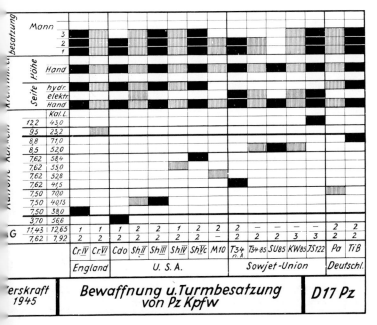
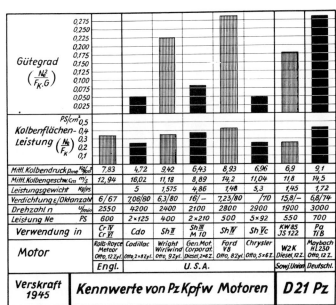

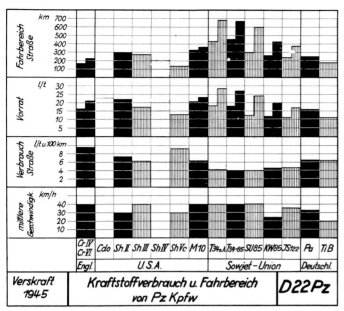
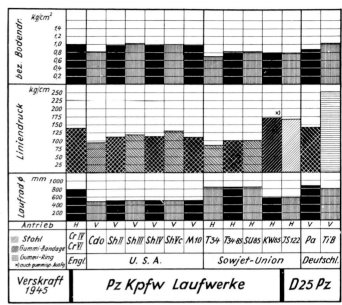
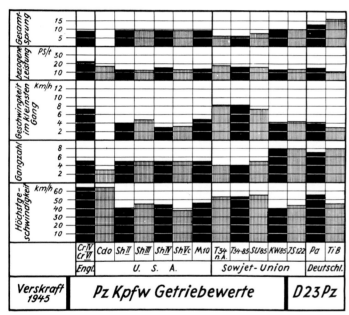
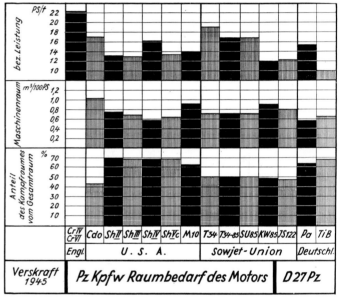
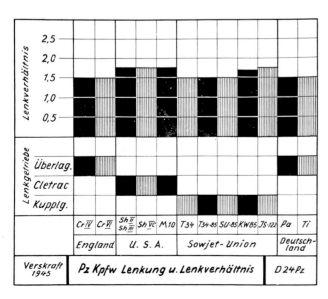
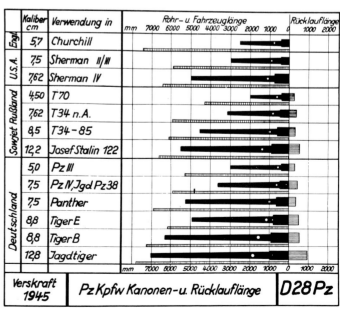

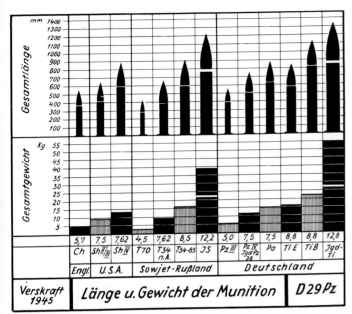

Verskraft 1945 — Länge u. Gewicht der Munition — D 29 Pz

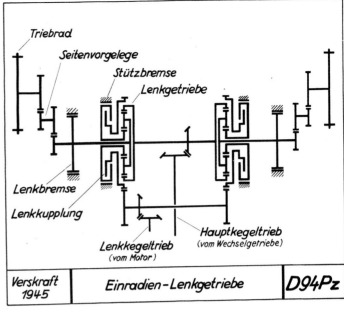

Verskraft 1945 — Einradien-Lenkgetriebe — D 94 Pz

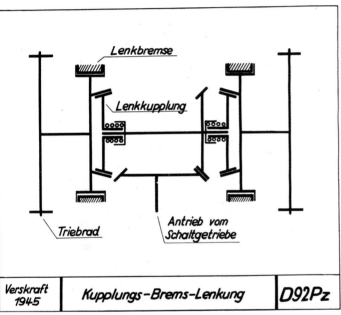

Verskraft 1945 — Kupplungs-Brems-Lenkung — D 92 Pz

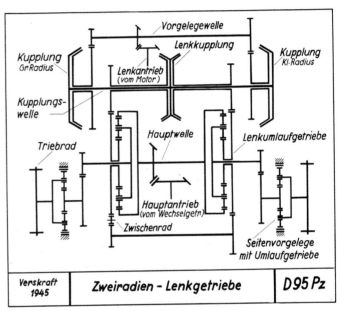

Verskraft 1945 — Zweiradien-Lenkgetriebe — D 95 Pz

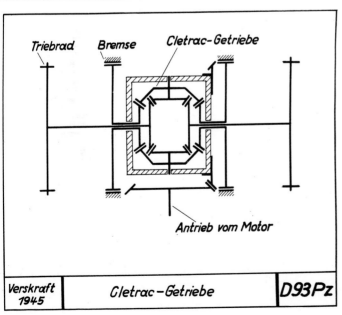

Verskraft 1945 — Cletrac-Getriebe — D 93 Pz

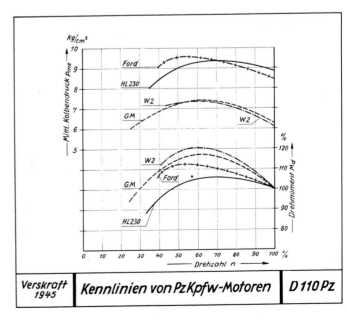

Verskraft 1945 — Kennlinien von PzKpfw-Motoren — D 110 Pz

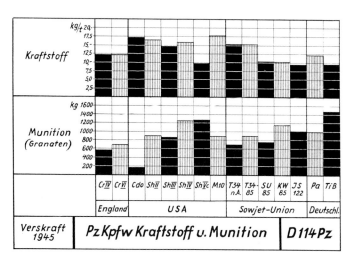

Appendix 20

Excerpts from the Pantherfibel

Vor Troja — wie's das Schicksal will —
Traf einst auf freiem Feld Achill
Die Königin der Amazonen.
Ein Zweikampf, denkt er, könnt sich lohnen.
Bewundernd, aber dennoch kalt,
Besieht er ihre Wohlgestalt.
Doch da kein Mensch ein ganzer Engel,
Drum hatte selbst die Fürstin Mängel.
Doch davon muß — um die zu sehn —
Man wie Achilles was verstehn.

Die Historie bringt uns Kleist in ernstem Gewande
Hier gekleistert jedoch sie als Histörchen erscheint.

Es gilt bei jedem Panzertyp
Genau das gleiche Grundprinzip.
Als erstes lernt schon der Rekrut:
Bleib klar im Kopf! Hab ruhig Blut!
Als zweites: Laß den Gegner ran,
Bis man ihn sicher knacken kann!
Doch kneift der Kerl, dann immer feste,
Dann rück ihm schleunigst auf die Weste.
Was ist's für einer? Schau gut hin!
Wann kann er Dich, wann kannst Du ihn?
Steckbrief und: Wo ist er zu packen?
Die schwachen Stellen mußt Du knacken!
Ein jeder Panzer ist zu brechen,
Kennst Du den Typ und seine Schwächen.

Dazu sieh Dir die Panzerbeschuß- und Erkennungstafeln an, die am Schluß der Fibel eingeheftet sind.

Die Fürstin bringt sich zum Duelle
Ein Mädchen mit für alle Fälle.
Achilles aber ebenfalls:
Ein Freund hält ihm die Maid vom Hals.
Achill, an Kraft drum unvermindert,
Noch von der Prothoe gehindert,
Rückt stolz dem königlichen Weibe
Wohl unterm Schild gedeckt zu Leibe.
Er hält zurück wie mit Reserven,
Behält vor allem seine Nerven.
Zeigt nie die Stelle, die verwündlich.
(Die Ferse ist bei ihm empfindlich.)
Nun hat Achill den Feind erreicht,
Daß er dem Schwert nicht mehr entweicht.
Du merkst für Deckung Dir dabei:
Von Wichtigkeit ist Zweierlei:

Du, Diomedes gehst
und schlägst die Frauen.
Ich bleib.   Achilles

Man bleibt in Deckung, bis man sich
Auch sagen kann: jetzt krieg ich dich!
Dann aber los, und mit Karbid!
So schießt man selbst, vor er was sieht.
Ein Angriff, zweitens, ist nichts nutz,
Geschieht er ohne Feuerschutz.
Denn Panzer sind meist nicht allein,
Sie kommen mindestens zu zwei'n.
Jedoch zwei treu verschworne Kumpeln
Kann man auch so nicht überrumpeln.
Vom andern wird gut aufgepaßt,
Bis Du den Gegner fertig hast.
Selbst jetzt zeig nur die Vorderfläche,
Und denk dabei an Siegfrieds Schwäche.
Des Gegners Schwächen nütz geschickt,
Damit Du ihn, nicht er Dich kriegt.
Wenn's geht, wird er zum Kampf gestellt,
Wo er's nicht für wahrscheinlich hält;
Denn jede Überrumpelung
Gibt Deinem Angriff Zeit und Schwung.

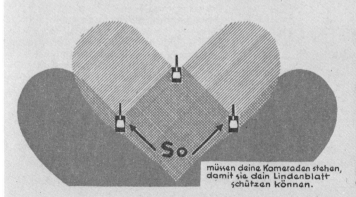

So müssen deine Kameraden stehen,
damit sie dein Lindenblatt
schützen können.

Index

Conception, Planning and Development of the Panther
Influence of the Russian T-34 on Panther development (July 1941). 10 ff
Contract to Rheinmetall-Borsig on 18 July 1941 for the development of a tank gun with high penetrating power. 11
Contract to Daimler-Benz and MAN on 25 November 1941 for the development of a tank with a 35 ton combat weight. 11
Contract for the development of a high-performance internal combustion engine of 700 metric hp by Maybach. 14
Development of a running gear with large wheel travel and effective damping. 18 ff
Decision of the Panther-Kommission on 11 May 1942 regarding introduction of the MAN Panther. 22 ff

Description of the Panther
High-performance 700 hp internal combustion engine 36 ff
Cyclone filter for engine combustion air 51 ff
Exhaust cooling 52, 125
Engine heating unit 160
Inertia-type starter 73 ff
7-gear all-synchro-mesh gearbox 55 ff
Clutch-and-brake steering unit 24
Replacement of the clutch-and-brake steering by the single-radius controlled differential steering 26
Solid-disk brakes 59 ff
Final drive with spur gearing 26
Automatic fire-extinguishing system 73 ff
Fighting compartment heating 160
Submersion 39
Winter equipment 160
Double torsion bars for 510 mm wheel travel 69 ff
Interleaved running gear 68 ff
Road wheels with replaceable rubber tires 68
Tracks with anti-skid grips 160
75mm Kampfwagenkanone (L/70) tank gun armament 75 ff
Ammunition 75
Ammunition performance 253
Hydraulic turret traverse drive 76 ff
Turret gunsight 77 ff
Armor protection 82 ff

Panther Production and Operations
Panther Ausf. D from November 1942 32 ff
Operations of the Panther in Operation "Zitadelle" (beginning of July 1943) 96 ff
Panther deficiencies due to series production without prior testing and troop trials 97
Panther Ausf. A 98 ff
Panther Ausf. G 125 ff
Panther Ausf. F 147 ff
Emergency program, conclusion of Panther production 155 ff
Cost and Material composition 254 ff
Proposal to manufacture the Panther in Italy 29, 86 ff

Panther Follow-on Developments, Testing and Experimenting
Air-cooled BMW 132 Dc radial aero engine 141
Water-cooled Daimler-Benz MB 507 diesel engine 108
Water-cooled Maybach HL 234 internal combustion fuel-injected engine 174 ff
Air and water-cooled diesel engines 174 ff
Hydrodynamic transmission 87 ff
Hydrostatic constant steering unit 89 ff
Electrically operated transmission 162, 172
Dual-radius steering unit 122
Final drive with planetary reduction 57, 60, 118
5- or 6-gear gearbox 153 ff
Conical disc spring suspension 156 ff
Resilient all-steel road wheels 142 ff
Staggered running gear 173
Reduction of armor alloy steel plating 144
Sleeve bearings in place of ball and roller bearings 146
Installation of the 75mm KwK (L/100) 29
Installation of the 88mm Pak 43/3 (L/71) 152
Air filtration system 146
Improved turret (Schmalturm) 147 ff
Automatic ammunition loader 153
Stabilized gunsight 154 ff
Stereoscopic range-finder 151 ff
Panther in foreign service (Italy, Hungary, Japan, France, USSR) 86 ff
Panther II 88, 169 ff
Panther E 50 156 ff
Development of night vision devices 164 ff

Panther Command and Observation vehicles 177 ff
Communications equipment 177
Armament 177

Jagdpanther 185 ff
88mm Pank 43/3 (L/71) 190
Ammunition performance 253
Production 196

Jagdpanther Follow-on Developments
128mm PaK 80 (L/55) installation
Recoilless installation of the main armament 197 ff

Bergepanther 200 ff
40 ton cable winch 204 ff
Production 212 ff

Flakpanzer (anti-aircraft tank) development on the Panther chassis 214 ff
Self-propelled Artillery (Armored Howitzers) - Development on the Panther Chassis 220 ff

Bibliography

Sources:

Willi A. Boelcke, *Deutschlands Rüstung im Zweiten Weltkrieg*
Duncan Crow/Robert J. Icks, *Encyclopedia of Tanks*
Eberhard Fechner, *Die Waffen des Kampfpanzers in Vergangenheit, Gegenwart und Zukunft*
H. Gaertner, *Die Bedeutung der infraroten Strahlen für militärische Verwendungszwecke*
Heinz Guderian, *Erinnerungen eines Soldaten*
Robert. J. Icks, *Tanks and armored vehicles*
Janusz Magnuski, Wozy Bojowe
F.W. von Mellenthin, *Panzer Battles*
Oskar Munzel, *Die deutschen gepanzerten Truppen bis 1945*
Walther Nehring, *Die Geschichte der deutschen Panzerwaffe 1916-1945*
Werner Oswald, *Die Kraftfahrzeuge und Panzer der Reichswehr, Wehrmacht und Bundeswehr*
Walter Rau, *Panzerungen*
Norbert Schausberger, *Rüstung in Osterreich 1938-1945*
H. Scheibert/C. Wagener, *Die deutschen Panzertruppe 1939-1945*
F.M. von Senger und Etterlin, *Die deutschen Panzer 1926-1945*
Walter J. Spielberger/Uwe Feist, *Armor Series 1-10*
Walter J. Spielberger, *Panzerkampfwagen Panther and its variations*
Walter J. Spielberger, *Die Kraftfahrzeuge und Panzer des österreichischen Heeres*
Rolf Stoves, *Die 1. Panzer-Division*
Johannes Thiede, *Federung und Dämpfung schneller Strassenfahrzeuge*
Friedrich Wiener, *Gepanzert auf Strasse und Schiene*
Heinrich Wüst, *Kraftfahrzeugtechnik des Kampfpanzers*

D 655/1 Gerätbeschreibung und Bedienungsanweisung zum Fahrgestell
D 655/2 Gerätbeschreibung und Bedienungsanweisung zum Turm
D 655/3 Beladeplan
D 655/4b Fristenplan für Schmier- und Pflegearbeiten
D 655/5 Handbuch für den Panzerfahrer
D 655/27 Panther-Fibel
D 655/30a Instandsetzungsanleitung für Panzerwarte, Laufwerk
D 655/30b Instandsetzungsanleitung für Panzerwarte, Triebwerk
D 655/30c Instandsetzungsanleitung für Panzerwarte, Motor
D 655/31a Werkstatthandbuch, Laufwerk
D 655/31b Werkstatthandbuch, Triebwerk
D 655/31c Werkstatthandbuch, Motor
D 655/60 Begleitheft
D 674/170 Gerätbeschreibung für Sonderwerkzeug HDv
428/1 Verzeichnis der Sonderwerkzeugsätze
D 2003 75mm Kampfwagenkanone 42 (L/70)
D 2030 88mm Panzerjägerkanone 43/2 (L/71)

Abbreviations

(selected abbreviations used in text and diagrams)
a/A alte Art, (old model)
A (2) Infanterieabteilung des Kriegsministeriums (Infantry Section of the War Ministry)
A (4) Feldartillerieabteilung des Kriegsministeriums (Field Artillery Section of the War Ministry)
A (5) Fussartillerieabteilung des Kriegsministeriums (Foot/Infantry Artillery Section of the War Ministry).
A (7) V Verkehrsabteilung des Kriegsministeriums (Transportation Section of the War Ministry)
AD (2) Allgemeines Kriegsdepartment, Abteilung 2 (Infanterie) (General War Department, Section 2 (Infantry))
AD (4) Allgemeines Kriegsdepartment, Abteilung 4 (Feldartillerie) (General War Department, Section 4 (Field Artillery))
AD (5) Allgemeines Kriegsdepartment, Abteilung 5 (Fussartillerie) (General War Department, Section 5 (Foot Artillery))
a.D. ausser Dienst (retired)
AG Aktiengesellschaft (joint-stock company)
Ah. Anhänger (trailer)
AHA/Ag K Allgemeines Heeresamt, Amtsgruppe Kraftfahrwesen (General War Office, Vehicle Affairs Group)
AK Artillerie-Konstruktionsbüro (Artillery Design Bureau)
AOK Armee-Oberkommando (High Command of an Army)
APK Artillerieprüfungskommission (Artillery Testing Commission)
ARW Achtradwagen (eight-wheeled vehicle)
Ausf. Ausführung (model)
BAK Ballon-Abwehr-Kanone (anti-balloon gun)
BMW Bayerische Motoren Werke (company name)
(D) Deutschland (German make)
DB Daimler-Benz (company name)
Dipl.-Ing. Diplom-Ingenieur (graduate engineer)
Dr.-Ing. Doktor der Ingenieurwissenschaften (Doctor of Engineering Sciences)
Dr. habil. Doktor habilitatus (doctor habilitatus)
E Entwicklung (developmental)
E- Einheits- (standard)
Fa Feldartillerie (field artillery)
FAMO Fahrzeug- und Motorenbau GmbH (company name)
Flak Flugabwehrkanone (anti-aircraft gun)
Fu Funk (radio)
Fu Ger Funkgerät (radio set)
g geheim (secret)
g Kdos geheime Kommandosache (secret: Command Material)
g RS geheime Reichssache (secret: Reich Material)
gl geländegängig (all-terrain)
Gmbh Gesellschaft mit beschränkter Haftung (limited liability company)
GraW Granatwerfer (mortar)
Gw Geschützwagen (mobile gun platform)
H Techn. V Bl Heerestechnisches Verordnungsblatt (Army Technical Directive)
Hanomag Hannoversche Maschinenbau AG (company name)
HK Halbkette, Halbkettenfahrzeug (half-tracked vehicle)
HLGR Hohlladungsgranate (hollow-bore shell)
HWA Heereswaffenamt (Army Weapons Office)
I.D. Infanteriedivision (infantry division)
I.G. Infanteriegeschütz (infantry gun (heavy caliber))
In. Inspektion (Inspection)
In. 6 Inspektion des Kraftfahrwesens (Inspection of Vehicle Affairs)
k klein, kleiner, kleines (small)
K Kanone (cannon/gun)
KD Krupp-Daimler (joint company names)
K.D. Kavalleriedivision (cavalry division)
Kfz Kraftfahrzeug (motorized vehicle)
KM Kriegsministerium (War Ministry)
Kp Krupp (company name)
Krad Kraftrad (motorcycle)
KwK Kampfwagenkanone (tank gun)

L/ Kaliberlänge (caliber length)
le leicht (light)
leFH leichte Feldhaubitze (light field howitzer)
leFK leichte Feldkanone (light field gun)
le.I.G. leichtes Infanteriegeschütz (light infantry gun)
le.W.S. leichter Wehrmachtsschlepper (light Wehrmacht prime mover)
Lkw Lastkraftwagen (transport truck)
m mittel, mittlerer (medium)
MAN Maschinenfabrik Augsburg-Nürnberg (company name)
MIAG Mühlenbau- und Industrie AG (company name)
MG Maschinengewehr (machine gun)
MP Maschinenpistole (automatic/machine pistol)
MTW Mannschaftstransportwagen (troop transport vehicle)
Mun.Pz Munitionspanzer (armored ammunition carrier)
n/A neue Art (new model)
NAG Nationale Automobilgesellschaft (company name)
(o) handelsüblich (standard model)
Ob.d.H. Oberbefehlshaber des Heeres (Commander-in-Chief of the Army)
OKH Oberkommando des Heeres (Army High Command)
OKW Oberkommando der Wehrmacht (Wehrmacht High Command)
Pak Panzerabwehrkanone (anti-tank gun)
P.D. Panzerdivision (tank division)
Pf Pionierfahrzeug (engineering vehicle)
Pkw Personenkraftwagen (personnel vehicle)
PS Pferde-Stärke (horsepower (metric))
Pz.Bef.Wg. Panzerbefehlswagen (armored command vehicle)
Pz.Gr. Panzergranate (armor-piercing (AP) shell)
Pz.Jg. Panzerjäger (tank destroyer)
Pz.Kpfw. Panzerkampfwagen (tank)
Pz.Sfl. Panzer-Selbstfahrlafette (armored self-propelled)
Pz.Spwg. Panzer-Spähwagen (armored reconnaissance vehicle)
(R) Raupen (tracked/caterpillar)
RhB Rheinmetall-Borsig (company name)
R/R Räder/Raupenantrieb (wheeled-cum-tracked drive)
RS Raupenschlepper (caterpillar tractor)
RSG Gebirgsraupenschlepper (caterpillar tractor for mountainous terrain)
s schwer (heavy)
Sankra Sanitätskraftwagen (ambulance)
schf. schwimmfähig (amphibious)
schg. schienengängig (on rails)
Sd.Kfz. Sonderkraftwagen (special purpose vehicle)
sFH schwere Feldhaubitze (heavy field howitzer)
Sf. Selbstfahrlafette (self-propelled)
Sfl. Selbstfahrlafette (self-propelled)
SmK Spitzgeschoss mit Stahlkern (pointed shell with hardened steel core)
SPW Schützenpanzerwagen (armored personnel carrier)
SSW Siemens-Schuckert Werke (company name)
StuG Sturmgeschütz (assault gun)
StuH Sturmhaubitze (assault howitzer)
StuK Sturmkanone (assault gun/cannon)
s.W.S. schwerer Wehrmachtsschlepper (heavy Wehrmacht prime mover)
(t) Tschechoslowakei (Czechoslovakia)
Tp Tropenausführung (tropical model)
TZF Turmzielfernrohr (turret gunsight)
ve voll entstört (fully shielded)
V- Versuchs- (test/prototype)
VFW Flakversuchswagen (test mobile anti-aircraft platform)
VK Vollkettenfahrzeug (fully-tracked vehicle)
VPK Verkehrstechnische Prüfungskommission (Transportation Sciences Testing Commission)
WaPrüf Waffenprüfungsamt (Office of Weapons Testing)
wg wassergängig (amphibious)
ZF Zahnradfabrik Friedrichshafen (company name)
Zgkw Zugkraftwagen (half-track prime mover)

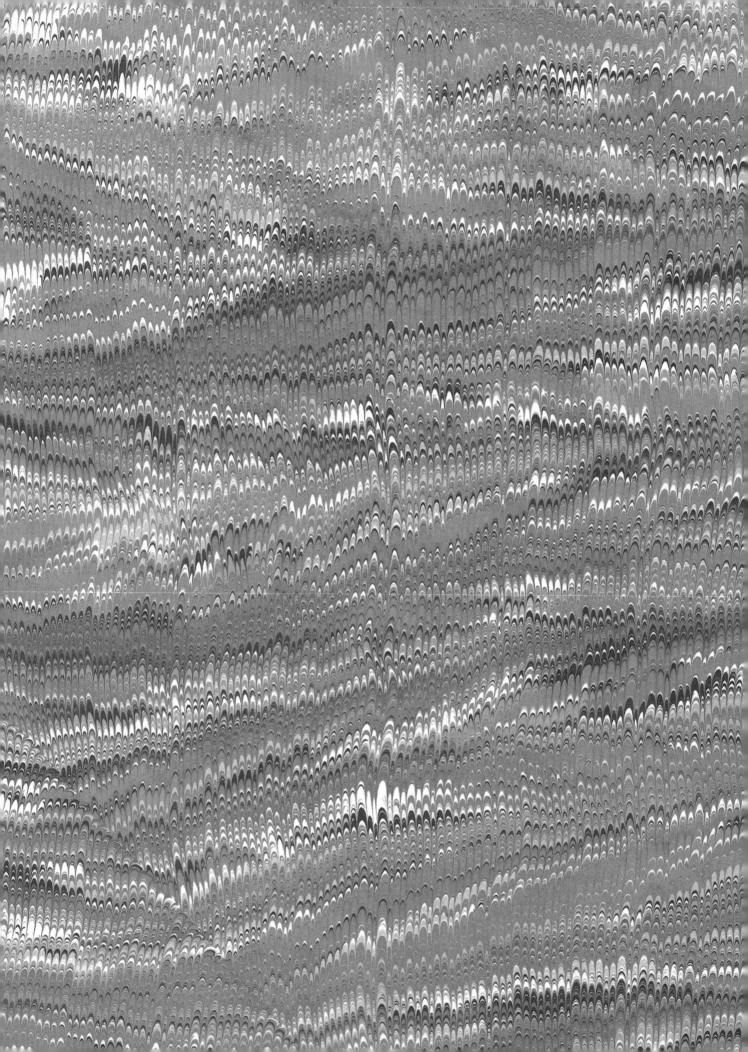